Simplified Design
of Filter Circuits

The EDN Series for Design Engineers

J. Lenk *Simplified Design of Filter Circuits*
C. Schroeder *Inside OrCAD Capture for Windows*
N. Kularatna *Power Electronics Design Handbook: Low-Power Components and Applications*
J. Lenk, *Simplified Design of Microprocessor-Supervisory Circuits*
C. Maxfield, *Designus Maximus Unleashed!*
EDN Design Ideas (CD-ROM)
C. Schroeder *Printed Circuit Board Design Using AutoCAD*
J. Lenk *Simplified Design of Voltage-Frequency Converters*
J. Lenk *Simplified Design of Data Converters*
F. Imdad-Haque *Inside PC Card: CardBus and PCMCIA Design*
C. Schroeder *Inside OrCAD*
J. Lenk *Simplified Design of IC Amplifiers*
J. Lenk *Simplified Design of Micropower and Battery Circuits*
J. Williams *The Art and Science of Analog Circuit Design*
J. Lenk *Simplified Design of Switching Power Supplies*
V. Lakshminarayanan *Electronic Circuit Design Ideas*
J. Lenk *Simplified Design of Linear Power Supplies*
M. Brown *Power Supply Cookbook*
B. Travis and I. Hickman *EDN Designer's Companion*
J. Dostal *Operational Amplifiers, Second Edition*
T. Williams *Circuit Designer's Companion*
R. Marston *Electronic Circuits Pocket Book: Passive and Discrete Circuits (Vol. 2)*
N. Dye and H. Granberg *Radio Frequency Transistors: Principles and Practical Applications*
Gates Energy Products *Rechargeable Batteries: Applications Handbook*
T. Williams *EMC for Product Designers*
J. Williams *Analog Circuit Design: Art, Science, and Personalities*
R. Pease *Troubleshooting Analog Circuits*
I. Hickman *Electronic Circuits, Systems and Standards*
R. Marston *Electronic Circuits Pocket Book: Linear ICs (Vol. 1)*
R. Marston *Integrated Circuit and Waveform Generator Handbook*
I. Sinclair *Passive Components: A User's Guide*

Simplified Design of Filter Circuits

John D. Lenk

Newnes
Boston   Oxford   Johannesburg   Melbourne   New Delhi   Singapore

Newnes is an imprint of Butterworth-Heinemann.

Copyright © 1999 by Butterworth-Heinemann

 A member of the Reed Elsevier group

All rights reserved.

No part of this publication may be reproduced, stored in a retrieval system, or transmitted in any form or by any means, electronic, mechanical, photocopying, recording, or otherwise, without the prior written permission of the publisher.

Recognizing the importance of preserving what has been written, Butterworth–Heinemann prints its books on acid-free paper whenever possible.

 Butterworth–Heinemann supports the efforts of American Forests and the Global ReLeaf program in its campaign for the betterment of trees, forests, and our environment.

Library of Congress Cataloging-in-Publication Data
Lenk, John D.
    Simplified design of filter circuits / John D. Lenk.
      p. cm.—(EDN series for design engineers)
    Includes index.
    ISBN 0-7506-9655-9 (pbk. : alk. paper)
    1. Electric filters—Design and construction. I. Title.
II. Series.
TK7872.F5L44 1999
621.3815′324—dc21   98-35333
                                                         CIP

British Library Cataloguing-in-Publication Data
A catalogue record for this book is available from the British Library.

The publisher offers special discounts on bulk orders of this book.
For information, please contact:
Manager of Special Sales
Butterworth-Heinemann
225 Wildwood Avenue
Woburn, MA 01801-2041
Tel: 781-904-2500
Fax: 781-904-2620

For information on all Butterworth-Heinemann publications available, contact our World Wide Web home page at: http://www.bh.com

10 9 8 7 6 5 4 3 2 1

Printed in the United States of America

Greetings from the Villa Buttercup!

To my wonderful wife, Irene: Thank you for being by my side all these years!
To my lovely family: Karen, Tom, Brandon, Justin, and Michael.
And to our Lambie and Suzzie: Be happy wherever you are!
And to my special readers: May good fortune find your doorway,
bringing good health and happy things. Thank you for buying my books!

To Karen Speerstra, Candy Hall, Duncan Enright, Pam Boiros, Katherine Greig, Pam Chester, Philip Shaw, Elizabeth McCarthy, the Newnes people, the UK people, and the EDN people: A special thanks for making me an international best seller, again (this is book 92).

Abundance!

Contents

Preface   xi
Acknowledgments   xiii

1 Introduction to Electronic Filters   1

1.1   Basic Filter Functions   1
1.2   Transfer Functions, Response, and Phase Shift   1
1.3   Filter Order, Poles, and Zeroes   3
1.4   Log Scales, –3 dB Frequencies, and Filter Q   4
1.5   Basic Filter Types   5
1.6   Filter Approximations   11
1.7   Classic Filter Functions   13
1.8   Passive Filters   18
1.9   Active Filters   19
1.10 Switched-Capacitor Filters   21
1.11 The Best Filter in the World   23

2 Typical Switched-Capacitor Filter   27

2.1   Basic Circuit Functions for MF10   27
2.2   Basic Filter Configurations   31
2.3   Design Hints for All Modes of Operation   31
2.4   MODE 1A: Non-Inverting Bandpass, Inverting Bandpass, and Lowpass   32
2.5   MODE 1: Notch, Bandpass, and Lowpass   32
2.6   MODE 2: Notch, Bandpass, and Lowpass   35
2.7   MODE 3: Highpass, Bandpass, and Lowpass   36
2.8   MODE 3A: Highpass, Bandpass, Lowpass, and Notch   37
2.9   MODE 4: Allpass, Bandpass, and Lowpass   39
2.10 MODE 5: Complex Zeros, Bandpass, and Lowpass   40
2.11 MODE 6A: Single-Pole, Highpass, and Lowpass   41
2.12 MODE 6B: Single-Pole Lowpass (Inverting and Non-Inverting)   42
2.13 Design Examples with the MF10   42

vii

3 Continuous (Active) Lowpass Filters 47

3.1 Basic Circuit Functions for MAX270/271 47
3.2 Programming the Cutoff Frequency 54
3.3 MAX270 Control Interface 58
3.4 MAX271 Control Interface 58
3.5 Digital Threshold Levels 59
3.6 Filter Performance 59
3.7 MAX271 Track-and-Hold 60
3.8 Power-Supply Configurations 60
3.9 Programming without a Microprocessor 62
3.10 Typical Application (Cascading) 63

4 Zero DC-Error Lowpass Filters 65

4.1 Basic Circuit Functions for MAX280/LTC1062 65
4.2 Using the Clock Divider Ratio 68
4.3 Using the Internal Oscillator 68
4.4 Using an External Clock 71
4.5 Choosing External Resistor and Capacitor Values 71
4.6 Input Voltage Range for MAX280 72
4.7 Internal Buffer 73
4.8 Filter Attenuation 73
4.9 Filter Noise 73
4.10 Transient Response for MAX280 73
4.11 Anti-Aliasing 74
4.12 LTC1062 Characteristics 75
4.13 Simplified Design Approaches (MAX280/LTC1062) 75

5 General-Purpose Lowpass Filters 91

5.1 Filter Response and Applications 91
5.2 Pin Descriptions for XR-1001/8 99

6 General-Purpose Elliptic Lowpass Filters 101

6.1 Filter Characteristics 101
6.2 Applications Requirements 106
6.3 Pin Descriptions for XR-1015/16 110

7 Tabular Design of Bandpass Filters 113

7.1 Using the Tables 113
7.2 Gain and Phase Relationships of the IC Filters 119
7.3 Cascading Non-Identical IC Filters 122
7.4 Cascading Identical IC Filters 125
7.5 Cascading Bandpass Filters in Mode 1 126

Contents **ix**

7.6	Cascading Bandpass Filters in Mode 2 127
7.7	Cascading More Than Two Identical 2nd-Order Sections 128

8   Practical Considerations for Switched-Capacitor Filters   129

8.1	SCFs versus Active RC Filters	129
8.2	Circuit Board Layout Problems	130
8.3	Power Supply Problems	132
8.4	Input Offset Problems	135
8.5	Slew Limiting	138
8.6	Aliasing in SCFs	139
8.7	Choosing a Filter Response	143
8.8	Noise in Filters (Active RC vs. SCF)	146
8.9	Clock Problems in SCFs	148
8.10	Bypass Capacitors for SCFs	150

9   Active RC Filters Using Current-Feedback Amplifiers   155

9.1	Basic CFA Operation	155
9.2	Active RC Filter Basics	155
9.3	Biasing CFA Active RC Filter Circuits	157
9.4	Highpass Active RC Filter	157
9.5	Lowpass Active RC Filter	159
9.6	Bandpass Active RC Filter	161
9.7	Bandpass Active RC Filter with Two CFAs	163
9.8	Bandpass Active RC Filter with Three CFAs (Bi-Quad)	165

10   Simplified Design Examples   169

10.1	DC-Accurate Filter for PLL Loops	169
10.2	Constant-Voltage Crossover Network	171
10.3	Infrasonic and Ultrasonic Filters	172
10.4	Lowpass Filters without DC Offset	173
10.5	Highpass Filter with Synthetic Inductor	174
10.6	DC-Accurate Notch Filter	176
10.7	Lowpass Filter for Anti-Aliasing	178
10.8	Communications Bandpass Filter with High Dynamic Range	180
10.9	Monolithic 5-Pole Lowpass Filter	181
10.10	Notch Filter Using an Op Amp as a Gyrator	184
10.11	Multiple Feedback Bandpass Filter	184
10.12	Filter with Both Notch and Bandpass Outputs	184
10.13	Lowpass Butterworth Active RC Filter	185
10.14	DC-Coupled Lowpass Active RC Filter	186
10.15	Universal State-Space Filter	187
10.16	Direct-Coupled Butterworth Filter	187

- 10.17 Bi-Quad Notch Filter   188
- 10.18 Elliptic Filter (Seven Section)   188
- 10.19 Basic Piezo-Ceramic-Based Filter   190
- 10.20 Piezo-Ceramic-Based Filter with Differential Network   190
- 10.21 Basic Crystal Filter   191
- 10.22 Single 2nd-Order Filter Section   192
- 10.23 Simple Highpass Butterworth Filter   192
- 10.24 Notch Filter with Adjustable Q   193
- 10.25 Easily-Tuned Notch Filter   194
- 10.26 Two-Stage Tuned Filter   194
- 10.27 Basic Tuned-Filter Circuit   194
- 10.28 Active RC Highpass Filter   195
- 10.29 Active RC Lowpass Filter   195
- 10.30 Notch Filter with High Q   196
- 10.31 Chebyshev Bandpass Filter   196
- 10.32 DC-Accurate Lowpass Bessel Filter   198
- 10.33 Simple Lowpass Filter   198
- 10.34 Wideband Highpass and Lowpass Filters   199
- 10.35 Fed-Forward Lowpass Filter   200
- 10.36 4.5-MHz Notch Filter   201
- 10.37 Spike Suppressor for Unregulated Power Supplies   201
- 10.38 DC-Accurate Lowpass/Bandpass Filter   201
- 10.39 Simple Bandpass Filter   201
- 10.40 Bandpass Filter with High Q 203
- 10.41 Chebyshev Bandpass Filter with Single Clock   203
- 10.42 Chebyshev Bandpass Filter with Two Clocks   204
- 10.43 Dual-Tracking 3-kHz Lowpass Filter   204
- 10.44 DC-Accurate Bessel Lowpass Filter   205
- 10.45 Filtering AC Signals from High DC Signals   207
- 10.46 Switched-Capacitor Filters (MAX291-97)   207
- 10.47 Tabular Design of Butterworth Lowpass Filters   209

For Further Information   211
Index   213

Preface

This book has something for everyone involved in electronics. Regardless of your skill level, the book shows you how to design and experiment with electronic filters.

For experimenters, students, and serious hobbyists, the book provides sufficient information to design and build filter circuits "from scratch." The design approach here is the same one used in all of the author's best-selling books on simplified and practical design.

Throughout the book, design problems start with guidelines for selecting all components on a trial-value basis. Then, using the guideline values in experimental circuits, the desired results (frequency range, tunability, offset, noise, passband and stopband responses, and so forth) are produced by varying the experimental components values, if needed.

Traditional filter design involves mathematics. Complex algebra is used extensively in filter-design equations and to represent filter functions. However, it is possible to design filters using simple equations. In many cases, filter design can be further simplified by means of tables, charts, and graphs. This is the approach taken in this book. *Math is kept to an absolute minimum.*

If you are a working engineer responsible for designing filters, or selecting IC filters, the variety of circuit configurations described here should generally simplify your task. Not only does the book describe filter-circuit designs, but it also covers the most popular forms of filter ICs available. Throughout the book, you will find a wealth of information on filter ICs and related components.

Chapter 1 is devoted to the basics of electronic filters. The information is included for those who are totally unfamiliar with filters, and for those who need a refresher. The descriptions here form the basis for understanding operation of the many ICs covered in the remaining chapters. Such an understanding is essential for simplified, practical design.

Chapter 2 describes simplified design with a typical universal switched-capacitor IC filter, and shows how the IC can provide all five basic filter types (bandpass, notch, lowpass, highpass, and allpass).

Chapter 3 describes simplified design with typical continuous (or active) IC filters, using simple tabular design.

Chapter 4 describes simplified design with simple switched-capacitor IC filters, using simple equations. Many practical examples are given in this chapter.

Chapter 5 describes simplified design with general-purpose lowpass IC filters.

Chapter 6 describes simplified design with general-purpose Elliptic lowpass filters.

Chapter 7 describes simplified (tabular) design of bandpass filters. The cascading of IC filters is also discussed throughout the chapter, using both identical and non-identical ICs.

Chapter 8 is devoted to the practical side of switched-capacitor filters, stressing both their capabilities and limitations.

Chapter 9 describes simplified design of active RC filters.

Chapter 10 describes simplified design for a cross section of filters. The circuits in this chapter can be used the way they are or, by scaling and altering component values, as a basis for simplified design of similar filters.

Acknowledgments

Many professionals have contributed to this book. I gratefully acknowledge their tremendous effort in making this work so comprehensive—it is an impossible job for one person. I thank all who have contributed, directly and indirectly.

I give special thanks to Alan Haun of Analog Devices, Syd Coppersmith of Dallas Semiconductor, Rose Hinejosa of EXAR Corporation, Jess Salter of GEC Plessey, Linda daCosta and John Allen of Harris Semiconductor, Ron Wenchfield of Linear Technology, David Fullagar and William Levin of Maxim Integrated Products, Fred Swymer of Microsemi Corporation, Linda Capcara of Motorola, Andrew Jenkins and Shantha Natrajan of National Semiconductor, Antonio Ortiz of Optical Electronics, Lawrence Fogel of Philips Semiconductors, John Marlow of Raytheon Company Semiconductor Division, Anthony Armstrong of Semtech Corporation, Ed Oxner and Robert Decker of Siliconix, Amy Sullivan of Texas Instruments, and Diane Freed Publishing Services.

I also thank Joseph A. Labok of Los Angeles Valley College for help and encouragement throughout the years.

Very special thanks to Karen Speerstra, Candy Hall, Duncan Enright, Pam Chester, Katherine Greig, Pam Boiros, Philip Shaw, Elizabeth McCarthy, Drew Bourn, Hillary Polk, Karen Burdick, Dawn Doucette, Laurie Hamilton, the Newnes people, the UK people, the EDN people, and the people of Butterworth-Heinemann for having so much confidence in me. I recognize that all books are a team effort and am thankful that I now work with the New First Team on this series.

And to Irene, my wife and super agent, I extend my thanks. Without her help, this book could not have been written.

CHAPTER 1

Introduction to Electronic Filters

This chapter is devoted to the basics of electronic filters. Although it is primarily for readers who are totally unfamiliar with filters, it is also very helpful for readers who need a refresher. As you may already know, traditional filter design involves mathematics. Complex algebra is used extensively in filter-design equations and to represent filter functions. However, it is possible to design filters using simple equations. In many cases, filter design can be further simplified by means of tables, charts, and graphs. This is the approach taken in this book. Math is kept to an absolute minimum. So before we get into the practical step-by-step details, let us establish some basic terms found in filter literature.

1.1 Basic Filter Functions

Figure 1–1 shows the basic filter function. A filter is an electrical network that alters the amplitude and/or phase characteristics of a signal with respect to frequency. Filters are used in electronic circuits to emphasize signals in certain frequency ranges and to reject signals in other frequency ranges. In this regard, filters have *gain,* which depends on signal frequency.

Using the simple example of Fig. 1–1, assume that a useful signal at frequency f1 is contaminated with an unwanted signal at f2. If the contaminated signal is passed through a filter that has very low gain at f2 compared to f1, the undesired signal can be removed and the useful signal will remain.

1.2 Transfer Functions, Response, and Phase Shift

Figure 1–2 shows a typical filter circuit. Figure 1–3 shows the amplitude and phase response of the filter. The response curves of Fig. 1–3 are similar to those of an amplifier. (If you are not familiar with amplifier response curves and how they are

2 SIMPLIFIED DESIGN OF FILTER CIRCUITS

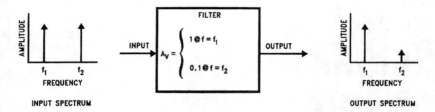

Figure 1–1. Basic filter functions. (National Semiconductor, *Linear Applications Handbook,* 1994, p. 1008)

obtained, read the author's *Simplified Design of IC Amplifiers,* Butterworth-Heinemann, 1996.)

Figure 1–2 also shows the equation for the filter *transfer function* or *network function* (which is the *ratio* of input and output signals). Note that the letter (s) is used to indicate complex frequency variables. This is necessary because the elements of the filter circuit respond differently to frequency changes. For example, the inductive reactance of the coil (or inductor) increases with frequency, while the capacitive reactance of the capacitor decreases as frequency increases. In turn, these reactances combine with the element resistance to form an impedance (which also changes with frequency).

For our purposes, the transfer-function magnitude (or input/output ratio) versus frequency is called the *amplitude response* (or the *frequency response,* especially if the filter is used in audio-frequency circuits). Similarly, the *phase response* of the filter gives the amount of *phase shift* introduced in sine-wave signals as a function of frequency. A change in phase of a signal also represents a change in time. As a result, the phase characteristics of a filter become especially important when dealing with complex signals where the time relationship between the signal components at different frequencies is critical.

The circuit shown in Fig. 1–2 is a *bandpass* filter because signals below a certain low frequency (f_l) and above a certain high frequency (f_h) are rejected or attenuated, but signals at frequencies between f_l and f_h are passed. The range of frequencies passed by a filter is known as the *passband.* (Note that some literature uses the terms band-pass, low-pass, and so on, as hyphenated words. The author prefers bandpass, lowpass, and so on, so you might find the terms used both ways in this book.)

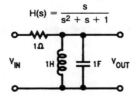

Figure 1–2.
Typical bandpass filter. (National Semiconductor, *Linear Applications Handbook,* 1994, p. 1009)

Introduction to Electronic Filters **3**

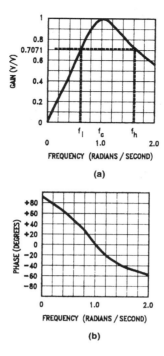

Figure 1-3.
Amplitude and phase response of bandpass filter. (National Semiconductor, *Linear Applications Handbook,* 1994, p. 1009)

1.3 Filter Order, Poles, and Zeroes

These terms appear quite often in filter literature, especially where the mathematical approach is used in design. We are interested in the simplified approach in this book, but you should have some idea where the terms fit into design.

The circuit of Fig. 1–2 is a *2nd order* filter. From a math standpoint, the *order* of a filter is the highest power of the variable (s) in the transfer function. (The transfer function in Fig. 1–2 shows s^2 of s to the power of 2.) From a simplified-design standpoint, the *order* of a filter is equal to the total number of capacitors and inductors (coils) in the circuit. (Note that a capacitor built by combining two or more individual capacitors counts as one capacitor.)

From a design standpoint (math or simplified) higher-order filters are more expensive because they require more components and are more complicated to design. However, the higher-order filters can more effectively discriminate between signals at different frequencies.

When two 2nd-order filters are *cascaded* (the output of one filter feeding directly into the input of a second identical filter) a 4th-order filter is formed. Figure 1–4 shows the effect on the frequency response when two 2nd-order *lowpass* filters are cascaded to form a 4th-order lowpass filter. Note the sharper *corner* and *slope* of the 4th-order filter in Fig. 1–4(b). Cascading is discussed further in Chapter 4.

4 SIMPLIFIED DESIGN OF FILTER CIRCUITS

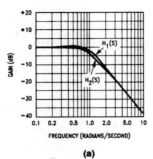

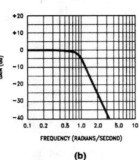

Figure 1–4.
Effect of cascading two lowpass filters. (National Semiconductor, *Linear Applications Handbook,* 1994, p. 1016)

Figure 1–5 is a *pole-zero* diagram or graphic representation of the transfer function shown in Fig. 1–2. To get such a diagram, the transfer function must first be factored to give:

$$H(s) = \frac{s}{\left(s + 0.5 + j\frac{\sqrt{3}}{2}\right)\left(s + 0.5 - j\frac{\sqrt{3}}{2}\right)}$$

Then the numerator of the factored equation is considered as a *zero* and is shown at the origin of the pole-zero diagram. The two denominators are called *poles*, with one at $2 = -0.5 - j\sqrt{3}/2$ and the other at $s = -0.5 + j\sqrt{3}/2$. (Now you see why we use the simplified, non-math, approach in this book!)

From a design standpoint, a pole anywhere to the right of the imaginary axis in Fig. 1–5 indicates instability. Stable filters will have all poles located on or to the left of the imaginary axis. (You may now forget about pole-zero diagrams!)

1.4 Log Scales, –3 dB Frequencies, and Filter Q

The scales shown in Fig. 1–3 are linear. As in the case of amplifier characteristics, the transfer and phase functions of filters generally use logarithmic (or log)

Figure 1-5.
Basic pole-zero diagram.
(National Semiconductor,
*Linear Applications
Handbook,* 1994, p. 1016)

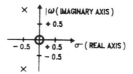

scales as shown in Fig. 1-4. This is because it is impractical to show a wide range of frequencies on linear scales (and log scales provide more symmetrical curves).

The *passband frequency limits* for all types of filters are usually assumed to be frequencies where the gain has dropped 3 decibels (3 dB) to 0.707 of the peak or maximum voltage gain. These are known as the *-3 dB frequencies* or the *cutoff frequencies.* Although the precise shape of a bandpass filter amplitude-response curve depends on the particular network, any 2nd-order bandpass response will have a peak value at the filter *center frequency,* which is the geometric mean of the -3 dB frequencies, or

$$f_c = \sqrt{f_l f_h}$$

where f_c is the center frequency, f_l is the lower -3 dB frequency, and f_h is the higher -3 dB frequency.

Another quantity used to describe filter performance is the filter Q. This is a measure of the "sharpness" of the amplitude response. For example, the Q of a bandpass filter is the ratio of the center frequency to the difference between the -3 dB frequencies (sometimes called the *-3 dB bandwidth*), and is calculated by $Q = f_c/(f_h - f_l)$. Figure 1-6 shows the response of various 2nd-order filters as a function of Q (with the gains and center frequencies normalized to unity).

From a practical standpoint, a filter with high Q has more reactance (either inductive or capacitive, or both) in relation to resistance in the circuits. Keep in mind that a high Q is not always a desirable characteristic in filters.

1.5 Basic Filter Types

As shown in Fig. 1-6, there are five basic filter types: bandpass, notch, lowpass, highpass, and allpass. The following is a summary of these types.

1.5.1 Bandpass Filters

Figure 1-2 shows the transfer function and circuit of a basic (passive) bandpass filter. Figure 1-7 shows the amplitude and phase response. Figure 1-8 shows some examples of bandpass-filter amplitude response.

The curve of Fig. 1-8(a) is the "ideal" bandpass response, with constant gain within the passband, zero gain outside the passband, and an abrupt boundary between the two. This response is impossible to get in practical filter circuits. Curves (b)

6 SIMPLIFIED DESIGN OF FILTER CIRCUITS

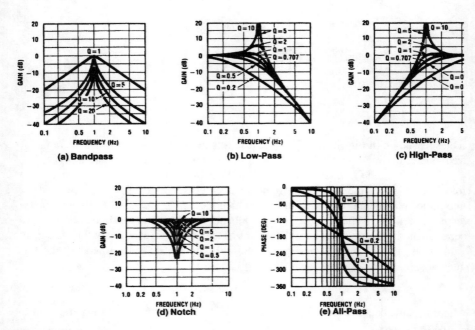

Figure 1-6. Filter response as a function of Q. (National Semiconductor, *Linear Applications Handbook*, 1994, p. 1015)

through (f) of Fig. 1-8 are examples of a few bandpass amplitude-response curves that approximate the ideal curve (with varying degrees of accuracy).

Note that some bandpass responses are very smooth, but others have *ripple* (gain variations) in the passbands, or in their *stopbands*. (The stopband is the range of frequencies over which unwanted signals are attenuated.) Bandpass filters have two

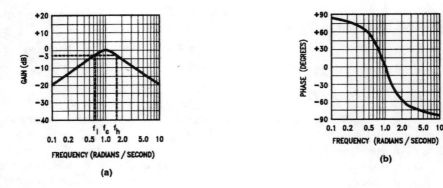

Figure 1-7. Amplitude and phase response of bandpass filters. (National Semiconductor, *Linear Applications Handbook*, 1994, p. 1010)

Introduction to Electronic Filters **7**

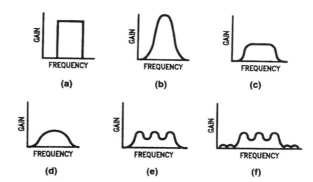

Figure 1-8.
Examples of bandpass-filter amplitude response. (National Semiconductor, *Linear Applications Handbook,* 1994, p. 1010)

stopbands, one above and one below the passband. The frequency at which attenuation begins is usually defined by system requirements. For example, a system specification might require that the signal must be attenuated at least 35 dB at 1.5 kHz. This defines the beginning of a stopband at 1.5 kHz.

The rate of change for attenuation between the passband and the stopband also differs with filter type, as does the slope of the curve. (Higher-order filters have steeper cutoff slopes.) Typically, the attenuation slope is expressed in dB/octave (an octave is a factor of 2 in frequency), or dB/decade (a decade is a factor of 10 in frequency).

Bandpass filters are used to separate a signal at one frequency, or within a band of frequencies, from signals at other frequencies. Using the example of Fig. 1–1, this function could be performed by a bandpass filter with a center frequency at f1, and f2 outside the passband. Depending on bandwidth, such a filter could also reject unwanted signals at other frequencies outside the passband.

1.5.2 Notch or Band-Reject Filters

Figure 1–9 shows the transfer function and circuit of a basic (passive) notch (or band-reject) filter. Figure 1–10 shows the amplitude and phase response. Figure 1–11 shows some examples of notch-filter amplitude response.

The curve of Fig. 1–11(a) is the ideal notch response. Notch filters are used to remove an unwanted frequency from a signal, while affecting all other frequencies as little as possible. For example, assume that an audio recording is contaminated with

Figure 1-9.
Transfer function and circuit of notch filter. (National Semiconductor, *Linear Applications Handbook,* 1994, p. 1011)

$$H_N(s) = \frac{V_{OUT}}{V_{IN}} = \frac{s^2 + 1}{s^2 + s + 1}$$

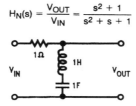

8 SIMPLIFIED DESIGN OF FILTER CIRCUITS

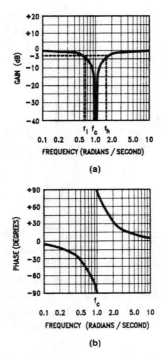

Figure 1-10.
Amplitude and phase response of notch filter. (National Semiconductor, *Linear Applications Handbook*, 1994, p. 1011)

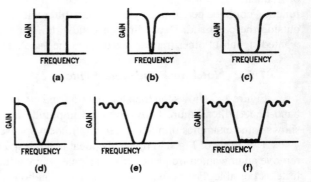

Figure 1-11.
Examples of notch-filter amplitude response. (National Semiconductor, *Linear Applications Handbook*, 1994, p. 1011)

60-Hz power-line hum. A notch filter with a center frequency of 60 Hz can remove the hum, but with little effect on the other audio signals.

1.5.3 Lowpass Filters

Figure 1–12 shows the transfer function and circuit of a basic (passive) lowpass filter. Figure 1–13 shows the amplitude and phase response. Figure 1–14 shows some examples of lowpass-filter response. A lowpass filter passes low-frequency signals

Introduction to Electronic Filters   9

Figure 1-12.
Transfer function and circuit of lowpass filter. (National Semiconductor, *Linear Applications Handbook,* 1994, p. 1012)

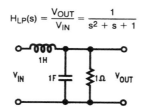

$$H_{LP}(s) = \frac{V_{OUT}}{V_{IN}} = \frac{1}{s^2 + s + 1}$$

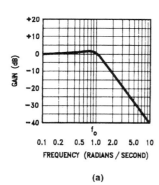

(a)

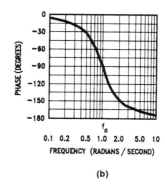

(b)

Figure 1-13. Amplitude and phase response of lowpass filter. (National Semiconductor, *Linear Applications Handbook,* 1994, p. 1012)

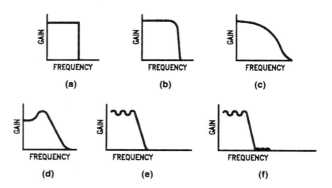

Figure 1-14.
Examples of lowpass-filter amplitude response. (National Semiconductor, *Linear Applications Handbook,* 1994, p. 1012)

and rejects signals at frequencies above the filter cutoff frequency. For this reason, lowpass filters are sometimes called *high-cut* filters.

1.5.4 Highpass Filters

Figure 1–15 shows the transfer function and circuit of a basic (passive) highpass filter. Figure 1–16 shows the amplitude and phase response. Figure 1–17 shows

10 SIMPLIFIED DESIGN OF FILTER CIRCUITS

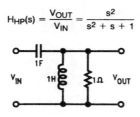

Figure 1–15.
Transfer function and circuit of highpass filter. (National Semiconductor, *Linear Applications Handbook,* 1994, p. 1013)

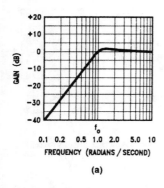

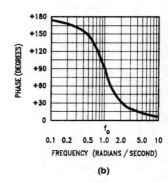

Figure 1–16. Amplitude and phase response of highpass filter. (National Semiconductor, *Linear Applications Handbook,* 1994, p. 1013)

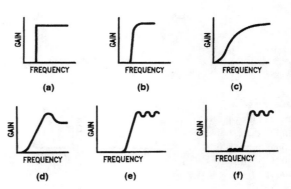

Figure 1–17.
Examples of highpass-filter amplitude response. (National Semiconductor, *Linear Applications Handbook,* 1994, p. 1013)

some examples of highpass-filter response. A highpass filter passes high-frequency signals and rejects signals at frequencies below the filter cutoff frequency. Highpass filters are sometimes called *low-cut* filters.

Highpass and lowpass filters are sometimes combined to separate signals of different frequencies. For example, the output of an audio amplifier can be passed to the tweeters of a stereo system through a highpass filter, and to the woofers through a lowpass filter, simultaneously. This is known as a *crossover* network.

1.5.5 Allpass or Phase-Shift Filters

The allpass filter shown in Fig. 1–6(e) is actually a phase-shift circuit. That is, the absolute value of the gain is equal to unity at all frequencies, but the phase changes as a function of frequency. Typically, such circuits are used to introduce phase shifts into signals and cancel or partially cancel any unwanted phase shift (previously imposed on the signals by other circuits or transmission devices). In this book, we are concerned primarily with circuits that separate signals of different frequencies.

1.6 Filter Approximations

Because the "ideal" curves shown Fig. 1–8, 1–11, 1–14, and 1–17 cannot be obtained in practical circuits, we must settle for the best *approximation* using some classic filter functions. Before we get into these functions, let us review the most important characteristics of filters from a simplified design standpoint.

1.6.1 Filter Order

As discussed, filter order is directly related to the number of components in the circuit, and therefore to the cost, physical size, and complexity. Higher-order filters are more expensive, take up more space, and are more difficult to design. The primary advantage of a higher-order filter is a steeper slope (or rolloff) than a lower-order filter.

1.6.2 Rolloff Rate

As discussed, the most common units for the amount of attenuation over a given ratio of frequencies are dB/octave and dB/decade. For simplified design, assume that the ultimate rolloff rate will be 20 dB/decade for *every filter pole* in the case of a lowpass or highpass filter. The rate is 20 dB/decade for *every pair of poles* for a bandpass filter. However, as is discussed throughout the following sections of this chapter, some filters have steeper attenuation slopes (rolloff rates) near the cutoff frequency than other filters of the same order.

1.6.3 Attenuation Rate Near Cutoff

If a filter is intended to reject a signal very close in frequency to a signal that must be passed, a *sharp cutoff characteristic* is required. Note that this steep slope might not continue to frequency extremes.

12 SIMPLIFIED DESIGN OF FILTER CIRCUITS

1.6.4 Transient Response

The amplitude curves described thus far show filter response to steady-state sine-wave signals. Because filters must operate with complex signals, it is often of interest to know how the filter will respond under transient conditions. As in the case of amplifier circuits, a step or pulse input can be applied to test the transient response of various filter types.

Figure 1–18 shows the responses of two lowpass filters to a step input. Curve (b) has a smooth reaction of the input step. Curve (a) shows some *ringing*. Curve (c) is the input signal. For simplified design, filters with sharper cutoff characteristics, or higher Q, will have more pronounced ringing.

1.6.5 Monotonicity

A filter has a monotonic amplitude response if the gain slope never changes sign; that is, if the gain always increases with increasing frequency or always decreases with increasing frequency. Lowpass and highpass filters can be truly monotonic as shown in Figs. 1–14(b) and (c) or 1–17(b) and (c). Bandpass filters can be monotonic on either side of the center frequency.

1.6.6 Passband Ripple

If a filter is not monotonic within the passband, the transfer function will show ripple as in Figs. 1–8(e) and (f), 1–11(f), 1–14(e) and (f), and 1–17(e) and (f). In some systems, the filter need not be monotonic, but the passband ripple must be limited to some maximum value (typically 1dB or less). Although bandpass and notch filters do not have true monotonic transfer functions, both filters can be free of ripple within their passbands.

1.6.7 Stopband Ripple

Some filter responses have ripple in the stopbands as shown in Figs. 1–8(f), 1–11(e) and (f), 1–14(f), and 1–17(f). From a practical standpoint, stopband ripple is not a problem as long as the signal to be rejected is sufficiently attenuated.

Figure 1–18.
Step responses of two lowpass filters. (National Semiconductor, *Linear Applications Handbook,* 1994, p. 1017)

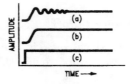

Introduction to Electronic Filters **13**

1.7 Classic Filter Functions

Most of the filters in common use follow one of the classic filter functions or approximations. These classic filters were developed by mathematicians (most bear their inventors' names), and each is designed to optimize some particular filter property. The following is a summary of the filter functions covered in this book.

1.7.1 Butterworth

Figure 1–19 shows the amplitude-response curves for Butterworth filters of various orders. Figure 1–20 shows the step responses for Butterworth lowpass filters. The Butterworth filter has a *maximally-flat* response (a nearly flat passband with no ripple).

The Butterworth rolloff is smooth and monotonic, with a lowpass or highpass rolloff rate of 20 dB/decade (6 dB/octave) for every pole. For example, a 5th-order Butterworth lowpass filter has an attenuation rate of 100 dB for every factor-of-10 increase in frequency (beyond the cutoff frequency).

1.7.2 Chebyshev

Figure 1–21 shows the amplitude-response curves for Chebyshev filters of various orders. Figures 1–21(a) and (b) show passband ripples of 0.1 dB and 0.5 dB,

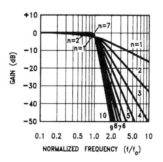

Figure 1–19.
Amplitude response for Butterworth filters of various orders. (National Semiconductor, *Linear Applications Handbook*, 1994, p. 1018)

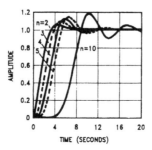

Figure 1–20.
Step responses for Butterworth lowpass filters. (National Semiconductor, *Linear Applications Handbook*, 1994, p. 1019)

14 SIMPLIFIED DESIGN OF FILTER CIRCUITS

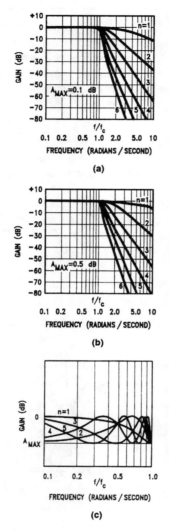

Figure 1–21.
Amplitude response for Chebyshev filters. (National Semiconductor, *Linear Applications Handbook*, 1994, p. 1020)

respectively. Figure 1–21(c) is an expanded view of the passband region. Note that the term A_{MAX} in Fig. 1–21 is the maximum allowable change in gain within the passband, and is also called *maximum passband ripple*. (The word "ripple" implies non-monotonic behavior, whereas A_{MAX} applies to monotonic response curves as well.) When A_{MAX} is used, you might also find the term A_{MIN}, which is the *minimum allowable attenuation* (referred to the maximum passband gain) within the stopband. These terms are shown in Fig. 1–22. The terms f_c and f_s are the cutoff frequency (or

Introduction to Electronic Filters   **15**

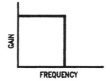

(a) "Ideal" Low-Pass Filter Response

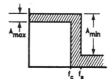

(b) Amplitude Response Limits
for a Practical Low-Pass Filter

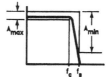

(c) Example of an Amplitude Response Curve Falling
with the Limits Set by f_c, f_s, A_{min}, and A_{max}

Figure 1-22.
Amplitude-response term
definitions. (National
Semiconductor, *Linear
Applications Handbook*,
1994, p. 1018)

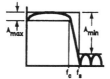

(d) Another Amplitude Response
Falling within the Desired Limits

passband limit) and the frequency at which the passband begins, respectively. Figure 1-23 shows the step response for Chebyshev lowpass filters.

The amount of passband ripple is one of the parameters used in specifying a Chebyshev filter. For example, a Chebyshev with an order of n will have n − 1 peaks or dips in the passband response. The nominal gain of the filter (unity gain in the case of the response in Fig. 1-21) is equal to the maximum passband gain. An odd-order Chebyshev will have a DC gain (in the lowpass case) equal to the normal gain, with "dips" in the amplitude-response curve equal to the ripple value. (This is why Chebyshev is said to have *equal ripple response*.)

An even-order Chebyshev lowpass will have the DC gain equal to the nominal filter gain *minus* the ripple value. The nominal gain for an even-order Chebyshev occurs at the peaks of the passband ripple. As a result, when designing a 4th-order Che-

Figure 1-23.
Step responses for Chebyshev lowpass filters. (National Semiconductor, *Linear Applications Handbook,* 1994, p. 1020)

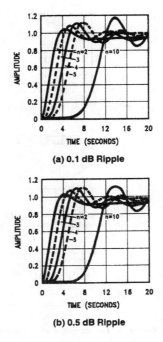

(a) 0.1 dB Ripple

(b) 0.5 dB Ripple

byshev lowpass filter with 0.5 dB ripple, and you want unity gain at DC, you must design for a nominal gain 0.5 dB.

The cutoff frequency of a Chebyshev is not assumed to be the –3 dB frequency as in the case of a Butterworth. Instead, the Chebyshev cutoff frequency is normally the frequency at which the ripple (or A_{MAX}) specification is exceeded.

The addition of passband ripple as a parameter makes the specification process for a Chebyshev filter more complicated than for a Butterworth, but also increases flexibility. When compared to the Butterworth, the Chebyshev has a steeper rolloff near the cutoff frequency. However, this is at the expense of monotonicity in the passband, and poorer transient response. (Compare Figs. 1–20 and 1–23.)

1.7.3 Bessel or Thompson

Figure 1–24 shows the amplitude-response curves for Bessel filters of various orders. Figure 1–25 shows the step response. The amplitude response is monotonic and smooth, but the Bessel cutoff characteristic is quite gradual compared to either the Butterworth or Chebyshev.

The important characteristic of a Bessel filter is an approximately *linear phase shift.* All filters show phase shift that varies with frequency. When the shift is not linear, the overall effect is to distort non-sine-wave signals. For example, Fig. 1–26 shows the response of a 4th-order Butterworth lowpass filter to a square-wave input.

Introduction to Electronic Filters **17**

Figure 1-24.
Amplitude response for Bessel filters. (National Semiconductor, *Linear Applications Handbook,* 1994, p. 1021)

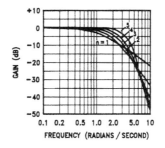

Figure 1-25.
Step response for Bessel lowpass filters. (National Semiconductor, *Linear Applications Handbook,* 1994, p. 1021)

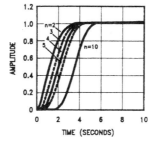

Figure 1-26.
Response of 4th-order Butterworth lowpass filter to square-wave input. (National Semiconductor, *Linear Applications Handbook,* 1994, p. 1021)

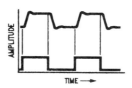

The ringing in the response shows that the nonlinear phase shift distorts the filtered square wave.

Figure 1-27 shows the response of a 4th-order Bessel lowpass filter to a square-wave input. Note the lack of ringing in the response. Except for the rounding of the corners, because of the reduction in high-frequency signal components, the response is a relatively undistorted version of the input square waves. Also notice the lack of ringing in the step response of Fig. 1-25, compared to the responses of Figs. 1-20 and 1-23.

18 SIMPLIFIED DESIGN OF FILTER CIRCUITS

Figure 1-27.
Response of 4th-order Bessel lowpass filter to square-wave input. (National Semiconductor, *Linear Applications Handbook,* 1994, p. 1021)

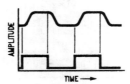

1.7.4 Elliptic

Figure 1–28 shows the amplitude-response curves for an elliptic filter (sometimes known as a Cauer filter). As shown, the cutoff slope is steeper than that of a Butterworth, Chebyshev, or Bessel, but the amplitude response has ripple in both the passband and the stopband, and the phase response is very nonlinear. However, if the primary concern is to pass frequencies falling within a certain frequency band and to reject frequencies outside that band, regardless of phase shifts or ringing, the elliptic response will perform that function with the lowest-order filter.

The elliptic function produces a sharp cutoff by adding notches in the stopband. These notches cause the transfer function to drop to zero at one or more frequencies in the stopband. Ripple is also introduced into the passband.

The elliptic filter function is usually specified by three parameters (excluding gain and cutoff frequency): passband ripple, stopband attenuation, and filter order n. (Because of the complexity, elliptic filter design usually requires a computer when the mathematical approach is used.)

1.8 Passive Filters

The filters discussed thus far (Figs. 1–2, 1–9, 1–12, and 1–15) are referred to as *passive* because they are made up of passive components (resistors, capacitors, and inductors). Because no amplifying elements (transistors, op amps, etc.) are used, pas-

Figure 1-28.
Amplitude response for elliptic filters. (National Semiconductor, *Linear Applications Handbook,* 1994, p. 1021)

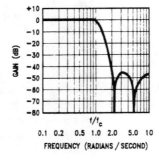

sive filters are the simplest to design. From a simplified design standpoint, you select the corresponding circuit for desired function (lowpass, highpass, etc.) and use standard equations to find resonant frequency of the circuit. This resonant frequency then becomes the *starting point* to determine the filter cutoff frequency (in the case of lowpass and highpass) or the center frequency (for bandpass and notch). You then test the completed circuit (using the amplifier test procedures discussed in Section 1.2) to find the filter response curve. If the curve is not as desired, you alter the component values as necessary. If you need the equations for resonant frequency, use the following simplified standbys:

$$F(MHz) = \frac{0.159}{L(\mu H) \times C(\mu F)}$$

$$L(\mu H) = \frac{2.54 \times 10^4}{F(kHz)^2 \times C(\mu F)}$$

$$C(\mu F) = \frac{2.54 \times 10^4}{F(kHz)^2 \times L(\mu F)}$$

Passive filters require no power supplies, are not restricted by the bandwidth limitations of op amps (so they can be used at very high frequencies), can handle larger current or voltage levels than active devices, and generate very little noise (only thermal noise). Passive filters also have some disadvantages: they provide no signal gain, input impedances are often lower than desired, output impedances are high, and buffer amplifiers might be required.

One of the big problems with passive filters is the selection of inductors. When high accuracy (1% or 2%), small physical size, or large inductance values are required, inductors are often prohibitively expensive. Standard values of inductors are not very closely spaced, and it is difficult to find an off-the-shelf inductor within 10 percent of any arbitrary value, so adjustable inductors are used. Tuning such inductors to the required values is time-consuming and expensive for larger quantities of filters.

Finally, when you go beyond 2nd-order passive filters, the simple equations do not apply and design becomes complex and time consuming. That is why we concentrate on *active filters* and *switched-capacitor filters* in this book.

1.9 Active Filters

Figure 1–29 shows a few of the commonly used active-filter circuits. Typically, active filters are made up of op amps with capacitors and resistors in the feedback loops to synthesize the desired filter characteristics. Inductors are not used, so the problems discussed in Section 1.8 do not apply. However, the problems of accuracy and value spacing for standard-value capacitors do apply, but to a lesser extent than for inductors.

20 SIMPLIFIED DESIGN OF FILTER CIRCUITS

The advantages and disadvantages of active filters are mostly the result of using op amps. For example, on the good side, active filters can provide high input impedance, low output impedance, and virtually any arbitrary gain. Active filters are generally easier to design than passive circuits (but not as easy as switched- capacitor filters). As to disadvantages, the performance of active filters at high frequencies is limited by the gain-bandwidth of the op amp, and circuit noise is determined by the op amp noise. (If noise is a problem, low-noise IC op amps must be used.)

1.9.1 Sallen-Key Active Filter

The 2nd-order Sallen-Key lowpass filter shown in Fig. 1–29(a) can be used as a building block for higher-order filters. By cascading two or more of these circuits, filters with orders of four or greater can be achieved. The two resistors and the two capacitors connected to the non-inverting op-amp input determine the filter cutoff frequency and affect the filter Q. The two resistors connected to the inverting input determine filter gain and Q. Because the components that determine both gain and cutoff frequency also affect Q, the gain and cutoff frequency cannot be changed independently.

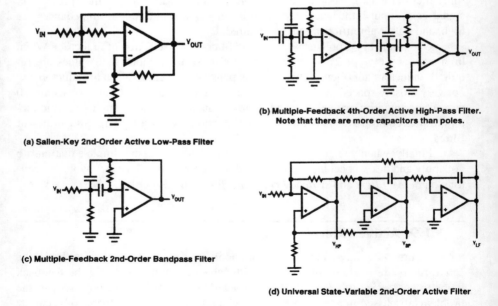

Figure 1–29. Commonly used active-filter circuits. (National Semiconductor, *Linear Applications Handbook,* 1994, p. 1023)

1.9.2 Multiple-Feedback Active Filters

The multiple-feedback filters shown in Figs. 1–29(b) and (c) use one op amp for each 2nd-order transfer function. Each highpass stage in the circuit of Fig. 1–29(b) requires three capacitors to get a 2nd-order response. As with the Sallen-Key, each component value affects more than one filter characteristic, so filter parameters cannot be independently adjusted.

1.9.3 Universal State-Variable Active Filters

The 2nd-order state-variable filter shown in Fig. 1–29(d) requires more op amps, but provides highpass, lowpass, and bandpass outputs from a single circuit. By combining the signals from the three outputs, any 2nd-order transfer can be realized.

1.9.4 Active-Filter Characteristics

When the center frequency is very low compared to the op-amp gain-bandwidth product, the characteristics of active RC filters depend primarily on external component tolerances and temperature drifts. For predictable results in critical filter circuits, external components with very good absolute accuracy and very low sensitivity to temperature variations must be used (and these can be expensive).

When the center frequency, multiplied by the filter Q, is more than a small fraction of the op-amp gain-bandwidth product, the filter response will deviate from the ideal transfer function. The degree of deviation depends on the filter circuit. (Some active-filter circuits minimize the effects of limited op-amp bandwidth, but not to the extent of a switched-capacitor filter.)

1.10 Switched-Capacitor Filters

Figure 1–30 shows the block diagram of a typical switched-capacitor filter. Figure 1–31 shows two IC versions of the filter (the National Semiconductor LMF40 and LMF60). Figure 1–32 shows the typical lowpass response of the IC filters. (The operation of switched-capacitor filters is discussed in Chapter 3.) As shown in Fig. 1–30, three filter functions (highpass, bandpass, and lowpass) can be obtained simultaneously with the addition of external resistors. Notch and allpass responses can also be obtained with different external-resistor connections. No capacitors or inductors are required. The filter center frequency is determined by an external clock. Accuracy is set by the clock, which is typically crystal controlled.

In effect, switched-capacitor filters are clocked, sampled-data systems, where the input signal to be filtered is sampled at a high rate and processed on a *discrete-time* basis rather than a continuous basis. This is the main difference between switched-capacitor filters and passive or active filters (which are sometimes called *continuous-time* filters).

22 SIMPLIFIED DESIGN OF FILTER CIRCUITS

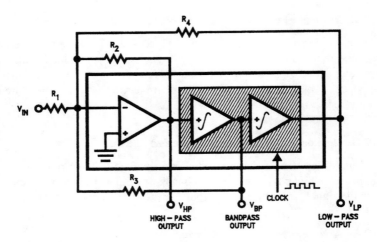

Figure 1-30. Typical switched-capacitor filter. (National Semiconductor, *Linear Applications Handbook,* 1994, p. 1024)

Because the switched-capacitor approach involves sampling, the Nyquist requirement is in effect. In simple terms, Nyquist states that the sampling frequency must be *greater than twice* the highest sampled frequency (and preferably higher).

The main advantage of switched-capacitor filters are accuracy (typically 0.2%), consistent and repeatable designs using inexpensive crystal-controlled oscilla-

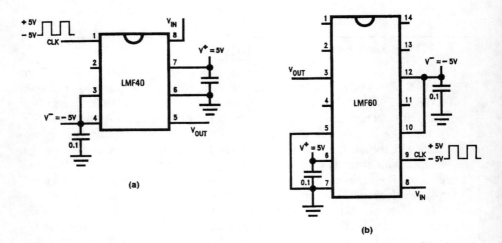

Figure 1-31. Typical external connections for IC switched-capacitor filters. (National Semiconductor, *Linear Applications Handbook,* 1994, p. 1024)

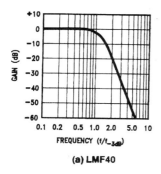
(a) LMF40

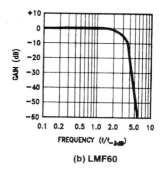
(b) LMF60

Figure 1–32. Typical lowpass response of IC filters. (National Semiconductor, *Linear Applications Handbook*, 1994, p. 1024)

tors, variable cutoff frequencies over a wide range by simply changing the clock, and low sensitivity to temperature changes. Switched-capacitor filters have one weakness: they produce more noise at their outputs (both random noise and clock feedthrough) than standard active filters.

Note that most IC switched-capacitor filters use some form of the universal, state-variable, active-filter configuration, such as shown in Fig. 1–29(d). However, in this book we reserve the term "active filter" for filters using non-switched, or continuous, active-filter techniques. Also, from a circuit standpoint, most switched-capacitor filters use non-inverting integrators, rather than the inverting integrators shown for conventional active filters in Fig. 1–29(d).

1.11 The Best Filter in the World

There is no best filter! Each filter technology offers a unique set of advantages and disadvantages that makes that approach a nearly ideal solution to some filter problem. The same approach can be totally unacceptable in other applications. Here is a quick look at the most important differences between passive, active, and switched-capacitor filters.

1.11.1 Accuracy

In most cases, the switched-capacitor filters have the advantage of better accuracy. Typical center-frequency accuracies are normally about 0.2 percent for most switched-capacitor ICs. Worst-case numbers range from 0.4 to 1.5 percent (assuming an accurate clock). To get this precision with passive or active filters, you must use very accurate resistors, capacitors, or inductors, or else the component values must be trimmed (which is costly and time consuming).

1.11.2 Cost

This is one area where there is no clear winner. If only a single-pole filter is needed, a passive RC or LC network will probably do the job. Also, there are some conventional active filters that can be built quite cheaply (especially where speed and accuracy are not required). Present-day switched-capacitor filters can also be inexpensive to buy, and take up very little expensive circuit-board space. When surface-mount is involved, passive components (especially capacitors) can be quite expensive.

1.11.3 Noise

Passive filters produce the least noise. Active filters produce more noise, but less than the typical switched-capacitor filter in IC form. The noise in passive and active filters is typically thermal noise. Switched-capacitor filters also have clock noise. Typical output levels from switched-capacitor filters are about 100 μV to 300 μV over a 20-kHz bandwidth. From a simplified-design standpoint, if the clock frequency is made high enough in relation to the signal frequency the clock can be ignored. Of course, if noise is critical, use the passive filter, or an active filter with very low noise op amps.

1.11.4 Offset Voltage

Passive filters have no inherent offset voltage. If a filter uses op amps and resistors (active filters) there is the usual offset caused by op-amp currents through the external resistors. Offset through the input resistors is multiplied by the op-amp gain. However, such offsets can be corrected using conventional techniques common to all IC op amps. (Refer to Section 1.2.)

Switched-capacitor filters have much larger offsets, ranging from a few millivolts to possibly 100mV (or larger in extreme cases). For these reasons, switched-capacitor filters are not recommended where DC precision is required.

1.11.5 Frequency Range

A single switched-capacitor filter can cover a center frequency range from 0.1 Hz or less to 100 kHz or more. A passive circuit or an op-amp active filter can be designed to operate at very low frequencies, but it will require very large (and probably expensive) reactive components. A fast op amp is necessary if an active filter is to work at 100 kHz and higher.

1.11.6 Tunability

Passive and active filters can be designed for virtually any center frequency. However, it is very difficult to vary that center frequency without changing the values of several components. A possible exception to this is a simple 2nd-order passive filter. Because the center (or cutoff) frequency of a switched-capacitor filter depends on the clock, it is easy to vary the frequency over a range of five to six decades with no change in external circuitry. This is an important advantage in applications that require multiple center frequencies.

1.11.7 Component Count and Circuit-Board Area

The switched-capacitor approach is the clear winner here. If the switched-capacitor filter is a dedicated, single-function IC, no external components are required (other than the clock). If the IC is programmable, typically four resistors per 2nd-order are required. Passive filters need a capacitor or inductor per pole. Active filters need at least one op amp, two resistors, and two capacitors per 2nd-order filter.

1.11.8 Aliasing

Switched-capacitor filters are sampled-data devices and are susceptible to aliasing when the input signal contains frequencies higher than one-half the clock frequency (the Nyquist requirement). Aliasing is also known as sampling error, and is the result of sampling at too slow a rate to determine the true picture. For example, if you sample a 100 Hz signal that is high (voltage, current, etc.) for 90 Hz and low for 10 Hz and you use a 10-Hz sampling rate, you might be taking the sample at only the 10 low points, and therefore assume that the entire 100-Hz signal is low.

Most switched-capacitor filters have clock-to-center frequency ratios of 50:1 to 100:1, so the frequencies at which aliasing begins are 25 to 50 times the center frequencies. When there are no signals with significant amplitudes at frequencies higher than one-half the clock frequency, aliasing will not be a problem. In lowpass or bandpass applications, the presence of signals at frequencies nearly as high as the clock rate will often be acceptable. Although these signals are aliased, they are reflected into the filter stopband and are thus attenuated by the filter.

Aliasing can sometimes be corrected by placing a simple RC filter ahead of the switched-capacitor filter to remove some of the unwanted high-frequency signals. This works best for both lowpass and bandpass switched-capacitor filters, but not for highpass (because the passive filter reduces the overall passband response).

1.11.9 Design Effort

With the possible exception of a simple 2nd-order passive filter, switched-capacitor filters are generally easier to use for both the experienced and inexperienced designer. (This is why we concentrate on switched-capacitor filters in this book!)

CHAPTER **2**

Typical Switched-Capacitor Filter

This chapter is devoted to simplified design with a typical switched-capacitor IC filter (the National Semiconductor MF10). The IC is chosen because it is universal and can provide all five basic filter types (bandpass, notch, lowpass, highpass, and allpass) as well as responses (Butterworth, Chebyshev, Bessel, and elliptic) by connecting a few external resistors and a clock. We start with a description of the basic circuit.

2.1 Basic Circuit Functions for MF10

Figure 2–1 shows the internal circuits in block form. The IC contains all of the elements necessary for two complete 2nd-order filters in one 20-pin package. Notice that each filter section contains two non-inverting integrators. The time constant of these integrators is controlled by the external clock. So before we go into circuit functions, let us review integrators.

2.1.1. Switched-Capacitor Integrators

Figure 2–2 shows the classic op-amp integrator circuit. Figure 2–3 shows how the resistor in Fig. 2–2 can be replaced with capacitors and switches. In the circuit of Fig. 2–2, the current flowing through feedback capacitor C is equal to VIN/R, and the circuit time constant is RC. In the circuit of Fig. 2–3, switches S1 and S2 are closed alternately by the clock. When switch S1 is closed (S2 open), capacitor C1 charges up to VIN. At the end of half a clock period, the charge on C1 is equal to VIN × C1. When the clock changes state, S1 opens and S2 closes. During this half of the clock period, all of the charge on C1 gets transferred to the feedback capacitor C2.

The amount of charge transferred from VIN to the summing junction (the inverting input) of the op amp during one complete clock period is VIN × C1. The current through C2 to the output is VIN × C1 × fCLK. The effective resistance from VIN

28 SIMPLIFIED DESIGN OF FILTER CIRCUITS

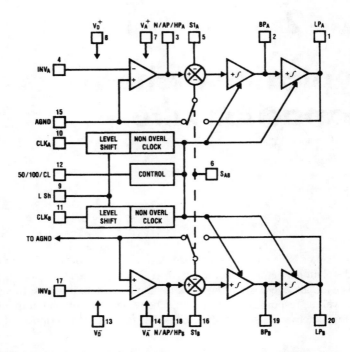

Figure 2-1. Internal circuits of MF10. (National Semiconductor, *Linear Applications Handbook,* 1994, p. 787)

to the inverting input is therefore R = VIN/1 = 1/(C1 × fCLK). In simple terms, S1, S2, and C1, when clocked as shown in Fig. 2–3, produce a clock-tunable time constant of C2/(C1 × fCLK).

The time constant (and frequency) of the switched-capacitor integrator depends on the ratio of two capacitors. Because both capacitors are fabricated on the same chip, the filter resonant frequency remains constant from IC to IC, and with changes in temperature. Also, the filter center frequency can be calculated directly from the clock frequency and a constant (such as fCLK/100, fCLK/50, and so on).

Figure 2-2.
Classic op-amp integrator. (National Semiconductor, *Linear Applications Handbook,* 1994, p. 796)

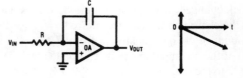

Figure 2-3.
Replacing integrator resistor. (National Semiconductor, *Linear Applications Handbook*, 1994, p. 796)

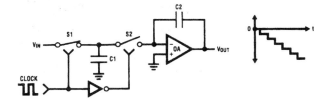

The actual integrators used in the IC are non-inverting as shown in Fig. 2–4. In this more elaborate circuit, S1A and S1B are closed simultaneously to charge C1 to VIN. The S2A and S2B are closed to connect C1 to the summing junction, but with the capacitor plates reversed to provide non-inverting operation. (If VIN is positive, VOUT moves positive when C2 acquires the charge from C1.)

2.1.2 Inputs/Outputs

As shown in Fig. 2–1, the outputs of each section of each filter are brought out to make the IC as universal as possible. There is a 3-input summing junction between the output of the summing op amp and the input of the first integrator. This makes it possible to subtract two inputs from the third.

From some filter configurations, one of the inverting inputs is brought out to serve as the signal input. The other inverting input is connected through an internal switch to either the lowpass output or analog ground, depending on the desired filter configuration. The direction of the input connection is common to both halves of the IC and is controlled by the voltage level on the SA/B input terminal. When tied to VD+ (the positive supply), the switch connects the lowpass output. When tied to VD– (the negative supply), the connection is to analog ground.

In other applications, it might be necessary to connect both inverting inputs of the summing junction for one half of the IC to ground, while the summing junction for the other half is connected to the lowpass output and ground. In such cases, connect the SA/B control to the negative supply, and make the connection to the lowpass output externally to the S1A (S1B) control pin.

Figure 2-4.
Basic non-inverting integrators. (National Semiconductor, *Linear Applications Handbook*, 1994, p. 796)

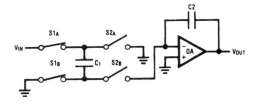

2.1.3 External Clock

A clock with close to 50-percent duty cycle is required to control the resonant frequency of the filter. Either TTL or CMOS clocks can be accommodated. If single-supply operation is desired, connect the level-shift (L Sh) control pin to ground.

A tri-level control pin (the 50/100/CL pin) sets the ratio of the clock frequency to the center frequency for both halves. When the pin is tied to V+, the center frequency is 1/50 of the clock frequency. When tied to mid-supply potential (ground, when powered from split supplies), the clock-to-center-frequency ratio is 100 to 1. When the pin is tied to V−, a power-saving current limiter shuts down operation and reduces the supply current by 70 percent.

Figure 2–5 shows a typical clock using a National Semiconductor COP452. This device can generate two independent 50-percent duty cycle clock frequencies. Each clock output is programmed by a 16-bit serial data word (N). This provides over 64,000 different clock frequencies for the filter from a single crystal.

2.1.4 Center Frequencies

The IC is intended for use with center frequencies up to 20 kHz, and is guaranteed to operate with clocks up to 1 MHz. Use the 50-to-1 clock for center frequencies greater than 10 kHz. The effect of using 100-to-1 or 50-to-1 clock-to-center-frequency ratios manifests itself in the number of "stair-steps" in the output waveform. This is shown in Fig. 2–6 where the 100:1 waveform is much smoother than the 50:1 waveform.

As discussed in Sec. 1.11.8, the aliasing problem can be minimized by increasing the clock (or sampling) frequency. However, increased clock frequencies result in increased DC offsets at the various filter outputs.

Several of the filter operating modes provide for altering the clock-to-center-frequency ratio by changing external resistor ratios. This can be used to get center frequencies of values other than 1/50 or 1/100 of the clock frequency. In multiple-stage, stagger-tuned filters, the center frequency of each stage can be set independently with resistors to allow the overall filter to be controlled by just one clock frequency.

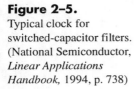

Figure 2–5.
Typical clock for switched-capacitor filters. (National Semiconductor, *Linear Applications Handbook*, 1994, p. 738)

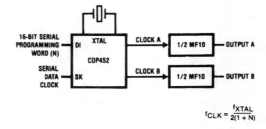

$$f_{CLK} = \frac{f_{XTAL}}{2(1 + N)}$$

Figure 2-6.
Sampled-data output waveforms. (National Semiconductor, *Linear Applications Handbook*, 1994, p. 788)

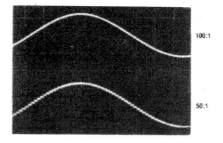

2.2 Basic Filter Configurations

As discussed in the remaining sections of this chapter, there are six basic configurations (or modes of operation) for the MF10. This makes it possible to realize a wide variety of filter responses. In all cases, no external capacitors are required. Design is established by selecting resistor ratios to set the desired passband gain and Q. The filter center frequency is set primarily by the external clock, but can be altered by selection of resistor ratios (in some modes of operation).

Each 2nd-order section of the IC can be treated in a modular fashion (with regard to individual center frequency, Q and gain) when cascading either the two sections within a package, or several packages for very high-order filters. This individuality is important when implementing the various response characteristics.

2.3 Design Hints for All Modes of Operation

The following is a general summary of design characteristics common to all modes of operation.

- The maximum supply voltage is ±7 V, or +14 V for single-supply operation. The minimum supply is 8 V.
- The maximum swing at any of the outputs is typically within 1 V of either supply.
- The internal op amps can source 3 mA and sink 1.5 mA. This is important when selecting a minimum resistor value.
- The maximum clock frequency is typically 1.5 MHz.

The Q and center-frequency (fo) relationship must be kept within reasonable limits. Here are some guidelines: For center frequencies less than 5 kHz, the fo × Q product can be as high as 300 kHz (Q must be less than or equal to 150). For example, a 3 kHz bandpass filter could have a Q as high as 100 with just one section. For center frequencies less than 20 kHz, limit the fo × Q product to 200 kHz. A 10-kHz bandpass design using a single section should have a Q no larger than 20.

Center-frequency matching from IC to IC for a given clock frequency is typically 0.2 percent. Center-frequency drift with temperature (excluding any clock-frequency drift) is typically ±10 ppm/°C with 50:1 switching and ±100 ppm/°C for 100:1.

The expressions for circuit dynamics given with each of the modes are important because they determine the voltage swing at each output as a function of circuit Q. A high-Q bandpass design can generate a significant peak in the response at the lowpass output at the center frequency.

Both sides of the IC are independent, except for supply voltages, analog ground, clock-to-center-frequency ratio setting, and internal switch setting for the three-input summing stage.

In the following descriptions of filter configurations (in this chapter), fo is the filter center frequency, Ho is the passband gain, and Q is the quality factor, and is equal to fo/BW, where BW is the –3-dB bandwidth measured at the bandpass output.

2.4 MODE 1A: Non-Inverting Bandpass, Inverting Bandpass, and Lowpass

Figure 2–7 shows the circuit connections for Mode 1A. Figure 2–8 shows the design equations and circuit dynamics. Mode 1A is the minimum external-component configuration (only two resistors) and is used for low-Q lowpass and bandpass applications. The non-inverting bandpass output (at pins 3 and 18) is for minimum-phase filter designs.

2.5 MODE 1: Notch, Bandpass, and Lowpass

Figure 2–9 shows the circuit connections for Mode 1. Figure 2–10 shows the design equations and circuit dynamics. Mode 1 is similar to Mode 1A, but with the addition of a resistor R1 between the input voltage (VIN) and the inverting input of

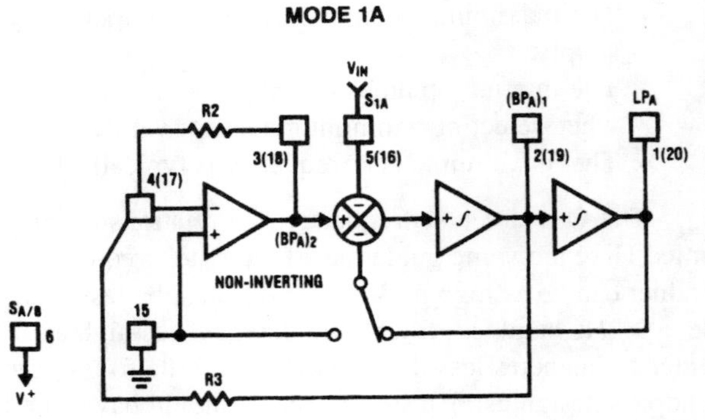

Figure 2–7.
Circuit connections for Mode 1A. (National Semiconductor, *Linear Applications Handbook,* 1994, p. 789)

Figure 2-8.
Design equations and circuit dynamics for Mode 1A. (National Semiconductor, *Linear Applications Handbook*, 1994, p. 789)

Design Equations

$$f_o = \frac{f_{CLK}}{100} \text{ or } \frac{f_{CLK}}{50}$$

$$Q = \frac{R3}{R2}$$

$$H_{OLP} = -1$$

$$H_{OBP_1} = -\frac{R3}{R2}$$

$$H_{OBP_2} = 1 \text{ (non-inverting)}$$

Circuit Dynamics

$H_{OBP_1} = -Q$ (this is the reason for the low Q recommendation)

$H_{OLP \text{ (peak)}} = Q \times H_{OLP}$

the op amp (at pins 4 or 17). Resistor R1 improves the output dynamics and allows bandpass designs with a much higher Q. The notch output features equal gain above and below the notch frequency.

2.5.1 Example of Mode 1 Filter

Figure 2–11 shows the MF10 connected to provide a 4th-order 1-kHz Butterworth lowpass filter. Both 2nd-order halves or sections of the IC are used to provide the 4th-order response (24 dB/octave or 80 dB/decade rolloff). This particular circuit is for single-supply operation, with the 10-V supply dropped to 5 V by the LM78L05 three-terminal regulator. Note that the analog-ground terminal (pin 15), the summer inputs S1A and S1B (pins 5 and 16), and the clock-switching control (pin 12) are biased to 5 V. These pins are grounded for symmetrical split-supply operation.

The input signal to be filtered is applied to the first half of the IC at pin 4 through an optional capacitor C and a 22-k resistor (R1 in Fig. 2–9). (Capacitor C is used if the input signal is not biased at 5V.) The output of the first 2nd-order filter is applied to the second half at pin 17 through a 10-k resistor. The output of the second 2nd-order filter is a 4th-order output at pin 20.

Figure 2-9.
Circuit connections for Mode 1. (National Semiconductor, *Linear Applications Handbook*, 1994, p. 789)

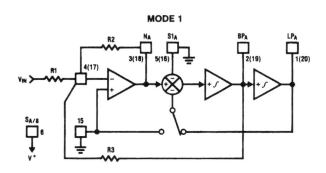

Design Equations

$$f_o = \frac{f_{CLK}}{100} \text{ or } \frac{f_{CLK}}{50}$$

$$f_{notch} = f_o$$

$$Q = \frac{R3}{R2}$$

$$H_{OLP} = -\frac{R2}{R1}$$

$$H_{OBP} = -\frac{R3}{R1}$$

$$H_{ON} = -\frac{R2}{R1} \text{ as } f \to 0 \text{ and as } f \to \frac{f_{CLK}}{2}$$

Circuit Dynamics

$H_{OBP} = H_{OLP} \times Q = H_{ON} \times Q$

$H_{OLP \text{ (peak)}} = Q \times H_{OLP}$ (if the DC gain of the LP output is too high, a high Q value could cause clipping at the lowpass output resulting in gain non-linearity and distortion at the bandpass output).

Figure 2-10. Design equations and circuit dynamics for Mode 1. (National Semiconductor, *Linear Applications Handbook*, 1994, p. 789)

There are six resistors in the circuit of Fig. 2–11. These correspond to resistors R1, R2, and R3 of Fig. 2–9, repeated for each half of the IC. Both the gain and Q of the filter halves are set by the resistors. For example, as shown by the equations of Fig. 2–10, the lowpass gain of the first half is set by R2/R1 (inverted). R2 corresponds to the 22-k resistor between pins 3 and 4. Thus, the gain of the first section is 22k/22k = 1 (or unity gain).

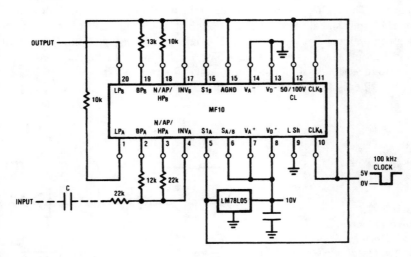

Figure 2-11. Example of 4th-order 1-kHz Butterworth filter. (National Semiconductor, *Linear Applications Handbook*, 1994, p. 793)

The gain of the second half is also set by R2/R1 (inverted). In the second half, R2 corresponds to the 10-k resistor between pins 17 and 18. R1 corresponds to the 10-k resistor between the output from the first half (pin 1) and the input to the second half (pin 17). This also provides a gain of 1 (10/10), so the overall filter response has unity gain. Because the signals have been inverted twice (once by each half), the output signal is back in phase with the input.

The Q for each section of the IC can be found using the equations of Fig. 2–10. As shown, Q is set by the ratio of R3/R2. For the first half of the circuit in Fig. 2–11, R3 corresponds to the 12-k resistor between pins 2 and 4 (feedback from the first integrator). R2 corresponds to the 22-k resistor between pins 3 and 4. (This is the same R2 that sets the lowpass gain.) Thus, the Q for the first section is 12k/22k, or 0.545.

For the second half of the Fig. 2–11 filter, R3 corresponds to the 13-k resistor between pins 17 and 19. R2 corresponds to the 10-k resistor between pins 17 and 18. As a result, the Q for the second section is 13k/10k, or 1.3. Notice that the resistors shown in Fig. 2–11 are typically 5 percent, so the precise Q and gain cannot be better than 5 percent (probably worse because more than one resistor is involved.)

Keep in mind that the calculations here provide *approximate resistor values* for a given Q or gain. The guidelines and tables given in remaining chapters provide more precise first-trial values (or you can use the software-math approaches, not found in any chapters of this book). No matter what approach you select, you still must adjust the trial values to get the exact response you want (or you have exceptional beginner's luck!).

The filter center frequency (fo) for both sections is found using the corresponding clock-to-center-frequency ratio equation of Fig. 2–1. Because we have a 100-kHz clock and want a 1-kHz center, we select the 100:1 ratio by connecting pin 12 to 5 V (which is one-half of the 10-V supply). If we had a 50-kHz clock, we could select the 50:1 ratio by connecting pin 12 to the 10-V supply. If we are using split supplies, ground pin 12 to get the 100:1 ratio.

We know how to select the resistor values and clock to produce a given center frequency, gain, and Q for a lowpass filter using Mode 1 of the MF10. The remainder of this chapter describes the circuit connections, design equations, and circuit dynamics for the remaining modes.

2.6 MODE 2: Notch, Bandpass, and Lowpass

Figure 2–12 shows the circuit connections for Mode 2. Figure 2–13 shows the design equations and circuit dynamics. Note that the notch output (pins 3 and 18) is useful in designing *elliptical highpass filters* because the notch frequency (fn) is less than the frequency of the center frequency (fo). The notch frequency is determined directly by the clock ratio. The center frequency is also determined by the clock ratio, but is altered by the ratio of R2 and R4. This allows tuning the clock-to-center frequency ratio to values greater than 100:1 or 50:1. Note that the high and low notch

36 SIMPLIFIED DESIGN OF FILTER CIRCUITS

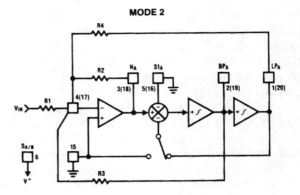

Figure 2-12.
Circuit connections for Mode 2. (National Semiconductor, *Linear Applications Handbook,* 1994, p. 790)

Design Equations

$$f_o = \frac{f_{CLK}}{100}\sqrt{1 + \frac{R2}{R4}} \text{ or } \frac{f_{CLK}}{50}\sqrt{1 + \frac{R2}{R4}}$$

$$f_n = \frac{f_{CLK}}{100} \text{ or } \frac{f_{CLK}}{50}$$

$$Q = \frac{\sqrt{R2/R4 + 1}}{R2/R3}$$

$$H_{OLP} = \frac{-\frac{R2}{R1}}{1 + \frac{R2}{R4}}$$

$$H_{OBP} = -\frac{R3}{R1}$$

$$H_{ON_1} \text{ (as } f \to 0) = \frac{-\frac{R2}{R1}}{1 + \frac{R2}{R4}}$$

$$H_{ON_2}\left(\text{as } f \to \frac{f_{CLK}}{2}\right) = -\frac{R2}{R1}$$

Circuit Dynamics

$$H_{OBP} = Q\sqrt{H_{OLP} \times H_{ON_2}} = Q\sqrt{H_{ON_1} \times H_{ON_2}}$$

Figure 2-13.
Design equations and circuit dynamics for Mode 2. (National Semiconductor, *Linear Applications Handbook,* 1994, p. 790)

outputs are set by the ratios of R1, R2, and R4. Filter Q is set by the ratios of R2, R3, and R4.

2.7 MODE 3: Highpass, Bandpass, and Lowpass

Figure 2–14 shows the circuit connections for Mode 3. Figure 2–15 shows the design equations and circuit dynamics. This mode is the switched-capacitor version of the classic state-variable filter shown in Fig. 2–16. In addition to requiring only

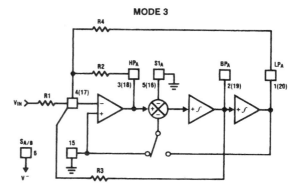

Figure 2-14. Circuit connections for Mode 3. (National Semiconductor, *Linear Applications Handbook*, 1994, p. 790)

Design Equations

$$f_o = \frac{f_{CLK}}{100}\sqrt{\frac{R2}{R4}} \text{ or } \frac{f_{CLK}}{50}\sqrt{\frac{R2}{R4}}$$

$$Q = \sqrt{\frac{R2}{R4}} \times \frac{R3}{R2}$$

$$H_{OHP} = -\frac{R2}{R1}$$

$$H_{OBP} = -\frac{R3}{R1}$$

$$H_{OLP} = -\frac{R4}{R1}$$

Circuit Dynamics

$$H_{OHP} = H_{OLP}\left(\frac{R2}{R4}\right)$$

$$H_{OLP\,(peak)} = Q \times H_{OLP}$$
$$H_{OBP} = Q\sqrt{H_{OHP} \times H_{OLP}}$$
$$H_{OHP\,(peak)} = Q \times H_{OHP}$$

Figure 2-15. Design equations and circuit dynamics for Mode 3. (National Semiconductor, *Linear Applications Handbook*, 1994, p. 790)

four external resistors, compared to many resistors and capacitors for the state-variable filter, the circuit of Fig. 2–14 is much easier to design. (Note the complexity of the equations in Fig. 2–16.)

Mode 3 is the most versatile mode of operation because the clock-to-center-frequency ratio can be externally tuned (by the ratios of R2 and R4) either above or below the 100:1 or 50:1 values. Mode 3 is best suited for multiple-stage Chebyshev filters controlled by a single clock.

2.8 MODE 3A: Highpass, Bandpass, Lowpass, and Notch

Figure 2–17 shows the circuit connections for Mode 3A. Figure 2–18 shows the design equations. A notch output is created from the circuit of Mode 3 by sum-

38 SIMPLIFIED DESIGN OF FILTER CIRCUITS

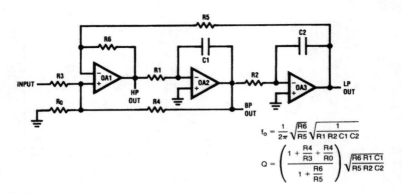

Figure 2-16. Classic universal state-variable 2nd-order active filter. (National Semiconductor, *Linear Applications Handbook,* 1994, p. 786)

ming the highpass and lowpass outputs through an external op amp. The ratio of the summing resistors Rh and Ri adjusts the notch frequency *independent* of the center frequency. This is useful in elliptic filter designs because the notch frequency can be shifted in relation to the center frequency as necessary to produce any steep dropoff required (within reason). Both frequencies are easily tuned by the clock and resistor ratios.

When cascading several stages of the MF10, the external op amp is needed only at the final output stage. The summing junction of the intermediate stages can be the inverting input of the MF10 internal op amp (pins 4 and 17).

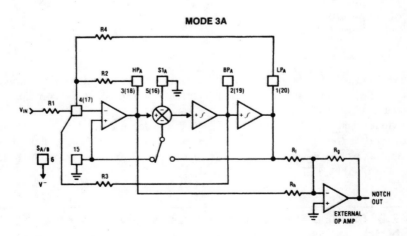

Figure 2-17. Circuit connections for Mode 3A. (National Semiconductor, *Linear Applications Handbook,* 1994, p. 791)

Design Equations

$$f_o = \frac{f_{CLK}}{100}\sqrt{\frac{R2}{R4}} \text{ or } \frac{f_{CLK}}{50}\sqrt{\frac{R2}{R4}}$$

$$Q = \sqrt{\frac{R2}{R4}} \times \frac{R3}{R2}$$

$$f_{notch} = \frac{f_{CLK}}{100}\sqrt{\frac{R_h}{R_l}} \text{ or } \frac{f_{CLK}}{50}\sqrt{\frac{R_h}{R_l}}$$

$$H_{OHP} = -\frac{R2}{R1}$$

$$H_{OLP} = -\frac{R4}{R1}$$

$$H_{OBP} = -\frac{R3}{R1}$$

$$H_{ON} \text{ (at } f = f_o) = \left| Q\left(\frac{R_g}{R_l}H_{OLP} - \frac{R_g}{R_h}H_{OHP}\right)\right|$$

$$H_{ONl} \text{ (as } f \to 0) = \frac{R_g}{R_l} \times H_{OLP}$$

$$H_{ONh}\left(\text{as } f \to \frac{f_{CLK}}{2}\right) = \frac{R_g}{R_h} \times H_{OHP}$$

Figure 2-18. Design equations and circuit dynamics for Mode 3A. (National Semiconductor, *Linear Applications Handbook*, 1994, p. 791)

2.9 MODE 4: Allpass, Bandpass, and Lowpass

Figure 2–19 shows the circuit connections for Mode 4. Figure 2–20 shows the design equations and circuit dynamics. In addition to providing the usual bandpass and lowpass functions, Mode 4 provides an allpass output at the S1A and S1B terminals (pins 3 and 18). The allpass output provides a *linear phase change with frequency* and results in a constant time delay. Note that Mode 4 restricts the gain at the allpass output to unity.

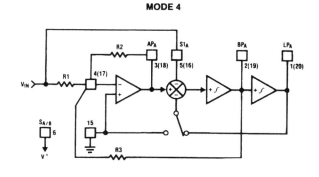

Figure 2-19. Circuit connections for Mode 4. (National Semiconductor, *Linear Applications Handbook*, 1994, p. 791)

Design Equations

$$f_o = \frac{f_{CLK}}{100} \text{ or } \frac{f_{CLK}}{50}$$

f_z (frequency of complex zero pair) $= f_o$

$$Q = \frac{R3}{R2}$$

Q_z (Q of complex zero pair) $= \frac{R3}{R1}$

$$H_{OAP} = -\frac{R2}{R1} = -1$$

$$H_{OLP} = -\left(\frac{R2}{R1} + 1\right) = -2$$

$$H_{OBP} = -\left(1 + \frac{R2}{R1}\right)\frac{R3}{R2} = -2\frac{R3}{R2}$$

Circuit Dynamics

$H_{OBP} = H_{OLP} \times Q = (H_{OAP} + 1)Q$

Figure 2-20.
Design equations and circuit dynamics for Mode 4. (National Semiconductor, *Linear Applications Handbook*, 1994, p. 791)

2.10 MODE 5: Complex Zeros, Bandpass, and Lowpass

Figure 2–21 shows the circuit connections for Mode 5. Figure 2–22 shows the design equations. Mode 5 features an improved allpass design over that of Mode 4. The Mode-5 circuit maintains a more constant amplitude with frequency at the complex zeros (C.z) output (pins 3 and 18). Also, the frequency of the pole pair and zero pair are resistor tunable.

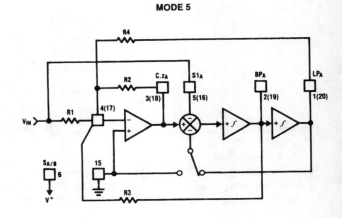

Figure 2-21.
Circuit connections for Mode 5. (National Semiconductor, *Linear Applications Handbook*, 1994, p. 792)

Design Equations

$$f_o = \frac{f_{CLK}}{100}\sqrt{1 + \frac{R2}{R4}} \text{ or } \frac{f_{CLK}}{50}\sqrt{1 + \frac{R2}{R4}}$$

$$f_z = \frac{f_{CLK}}{100}\sqrt{1 - \frac{R1}{R4}} \text{ or } \frac{f_{CLK}}{50}\sqrt{1 - \frac{R1}{R4}}$$

$$Q = \frac{R3}{R2}\sqrt{1 + \frac{R2}{R4}}$$

$$Q_z = \frac{R3}{R1}\sqrt{1 - \frac{R1}{R4}}$$

$$H_{O(C.z)} \text{ as } f \to 0 = \frac{R2(R4 - R1)}{R1(R2 + R4)}$$

$$H_{O(C.z)} \text{ as } f \to \frac{f_{CLK}}{2} = \frac{R2}{R1}$$

$$H_{OBP} = \frac{R3}{R2}\left(1 + \frac{R2}{R1}\right)$$

$$H_{OLP} = \frac{R4}{R1}\left(\frac{R2 + R1}{R2 + R4}\right)$$

Figure 2-22.
Design equations for Mode 5. (National Semiconductor, *Linear Applications Handbook*, 1994, p. 792)

2.11 MODE 6A: Single-Pole, Highpass, and Lowpass

Figure 2–23 shows the circuit connections for Mode 6A. Figure 2–24 shows the design equations. By using only one of the internal integrators, Mode 6A is useful for creating odd-ordered cascaded filter responses by providing a real pole that is clock-tunable to track the resonant frequency of other 2nd-order MF10 sections. The cutoff frequency (or corner frequency) is resistor-tunable.

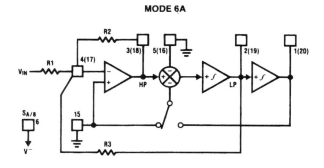

Figure 2-23.
Circuit connections for Mode 6A. (National Semiconductor, *Linear Applications Handbook*, 1994, p. 792)

Figure 2-24.
Design equations for Mode 6A. (National Semiconductor, *Linear Applications Handbook*, 1994, p. 792)

Design Equations

$$f_c \text{ (cut-off frequency)} = \frac{f_{CLK}}{100}\left(\frac{R2}{R3}\right) \text{ or } \frac{f_{CLK}}{50}\left(\frac{R2}{R3}\right)$$

$$H_{OLP} = -\frac{R3}{R1}$$

$$H_{OHP} = -\frac{R2}{R1}$$

2.12 MODE 6B: Single-Pole Lowpass (Inverting and Non-Inverting)

Figure 2-25 shows the circuit connections for Mode 6B. Figure 2-26 shows the design equations. Mode 6B uses only one of the integrators for a single-pole lowpass, and the input op amp as an inverting amplifier, to provide a non-inverting lowpass output. As in the case of 6A, Mode 6B is useful for designing odd-ordered lowpass filters.

2.13 Design Examples with the MF10

Figure 2-11 shows how the MF10 can be connected to provide a 4th-order 1-kHz Butterworth lowpass filter. The remainder of this section describes some additional examples of filter design using the MF10.

2.13.1 Input Filter and Sample/Hold

Figure 2-27 shows how the MF10 can be connected as an input filter and sample/hold (S/H) circuit for a data-acquisition system. This configuration provides bandlimiting (or anti-aliasing) and allows larger-amplitude, higher-frequency input signals. By gating off the applied clock, the switched-capacitor integrators will hold the last sampled voltage value. The drop rate of the output voltage during the hold time is approximately 0.1 mV per millisecond.

Figure 2-25.
Circuit connections for Mode 6B. (National Semiconductor, *Linear Applications Handbook*, 1994, p. 793)

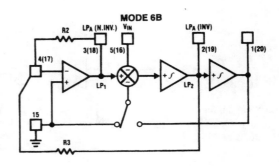

Figure 2-26.
Design equations for Mode 6B. (National Semiconductor, *Linear Applications Handbook*, 1994, p. 793)

Design Equations

$$f_c \text{ (cut-off frequency)} = \frac{f_{CLK}}{100}\left(\frac{R2}{R3}\right) \text{ or } \frac{f_{CLK}}{50}\left(\frac{R2}{R3}\right)$$

$$H_{OLP} \text{ (inverting output)} = -\frac{R3}{R2}$$

$$H_{OLP} \text{ (non-inverting output)} = +1$$

2.13.2 Quadrature Sine-Wave Generator

Figure 2–28 shows how the MF10 can be connected to generate stable-amplitude sine and cosine outputs without using automatic-gain-control (AGC) circuits. The MF10 operates as a Q-to-10 bandpass filter that rings at the resonant frequency in response to a step input change. The ringing signal is fed to an LM311 op amp that creates a square-wave input signal to the bandpass to sustain oscillation.

The bandpass output is the filtered fundamental frequency of a 50-percent duty-cycle square wave. A 90° phase-shifted signal of the same amplitude is available at the lowpass output through the second integrator of the MF10. The frequency of oscillation is set by the center frequency of the filter, as controlled by the clock and the 50:1/100:1 control (pin 12). The output amplitude is set by the peak-to-peak (p-p) swing of the square-wave input. (In this circuit, the p-p swing is defined by the back-to-back diode clamps at the LM311 output.)

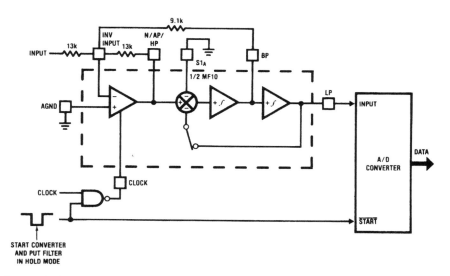

Figure 2–27. Input filter and sample/hold. (National Semiconductor, *Linear Applications Handbook*, 1994, p. 794)

44 SIMPLIFIED DESIGN OF FILTER CIRCUITS

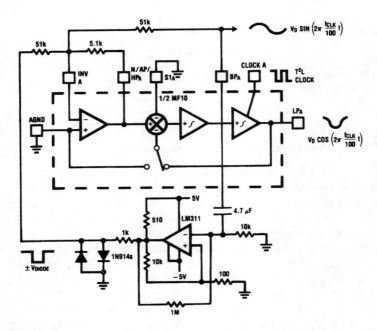

Figure 2–28. Quadrature sine-wave generator. (National Semiconductor, *Linear Applications Handbook,* 1994, p. 794)

2.13.3 Full-Duplex 300-Baud Modem Filter

Figure 2–29 shows how two MF10s can be connected to form a complete 300-baud, full-duplex modem filter. Because all four 2nd-order sections of the filters are used, the result is an 8th-order, 1-dB ripple Chebyshev bandpass that functions as both an 1170-Hz originate filter and a 2125-Hz answer filter. Control of answer or originate operation is set by the logic level at 50:1/100:1 inputs (pin 12) so that only one clock frequency is required. The overall filter gain is 22 dB.

Construction of this filter on a PC board is more compact than an RC active filter, and much more cost effective (considering the level of precision required). An even more attractive implementation from a space-savings standpoint would be a hybrid circuit. A film-resistor array connecting the two MF10 dies could produce the entire filter in one package. This would require only seven external connections for input, output, clock, supplies, and originate/answer control.

Typical Switched-Capacitor Filter 45

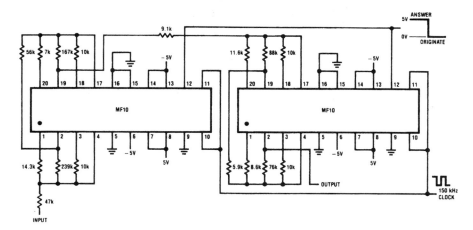

Figure 2-29. Full-duplex 300-baud modem filter. (National Semiconductor, *Linear Applications Handbook,* 1994, p. 795)

CHAPTER **3**

Continuous (Active) Lowpass Filters

This chapter is devoted to simplified design with typical continuous (or active) IC filters (the Maxim MAX270/271). These ICs are chosen because they provide lowpass filtering, with a wide range of cutoff frequencies, using simple tabular design. The ICs are digitally programmable, dual 2nd-order continuous-time (active) lowpass devices. Their typical dynamic range of 60 dB surpasses most switched-capacitor filters, which require additional filtering to remove clock noise. As discussed in Chapter 1, because there is no clock, there is no clock noise in continuous filters. Also, because continuous filters are not switched, there is no aliasing problem.

The two filter sections of the ICs are independently programmable by either microprocessor (μP) control or pin strapping. Cutoff frequencies in the 1-kHz to 25-kHz range can be selected with a simple table. The MAX270 has an on-board, uncommitted op amp. The MAX271 has an internal track-and-hold (T/H). Both filters are ideal for anti-aliasing and DAC (digital/analog converter) smoothing applications. The dual sections can be cascaded for higher-order responses. No external components are required.

3.1 Basic Circuit Functions for MAX270/271

Figure 3–1 shows the internal circuits in block form. Figures 3–2 and 3–3 show the MAX271 connected for P and pin-strap programming, respectively. Figure 3–4 shows a typical operating circuit for the MAX270. Figures 3–5 and 3–6 show the pin configurations for the MAX270 and MAX271, respectively. Figures 3–7, 3–8, and 3–9 show the electrical, timing, and operating characteristics, respectively. Figures 3–10 and 3–11 show the pin descriptions for the MAX270 and MAX271, respectively. Figure 2 referred to in Fig. 3–9 is the timing diagram of Fig. 3–12 in this book.

Both the MAX270 and MAX271 contain two independent, 2nd-order, Sallen-Key lowpass-filter sections (A and B) to provide a frequency-vs.-gain rolloff of about

48 SIMPLIFIED DESIGN OF FILTER CIRCUITS

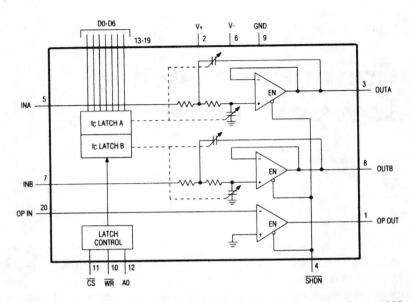

Figure 3–1. Internal circuits of MAX270. (Maxim New Releases Data Book, 1992, p. 6-68)

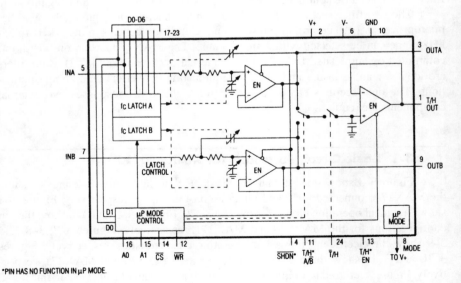

*PIN HAS NO FUNCTION IN µP MODE.

Figure 3–2. Internal circuits of MAX271 in µP mode. (Maxim New Releases Data Book, 1992, p. 6-68)

Continuous (Active) Lowpass Filters **49**

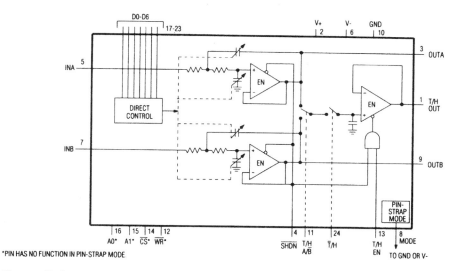

Figure 3-3. Internal circuits of MAX271 in pin-strip mode. (Maxim New Releases Data Book, 1992, p. 6-69)

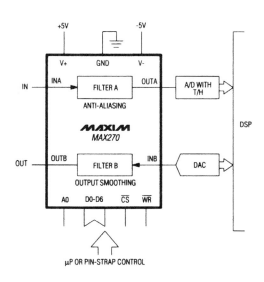

Figure 3-4.
Typical operating circuit for MAX270. (Maxim New Releases Data Book, 1992, p. 6-61)

50 SIMPLIFIED DESIGN OF FILTER CIRCUITS

Figure 3–5.
Pin configurations for
MAX270. (Maxim New
Releases Data Book,
1992, p. 6-61)

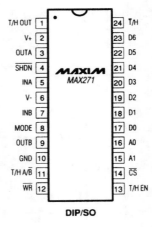

Figure 3–6.
Pin configurations for
MAX271. (Maxim New
Releases Data Book,
1992, p. 6-75)

40dB/decade. These ICs are not switched-capacitor filters, but have a continuous-time design similar to discrete active filters using op amps (similar to that shown in Fig. 2–16).

ABSOLUTE MAXIMUM RATINGS

V+ to V- .... -0.3V, +17V
V+ to GND .... -0.3V, +8.5V
V- to GND .... +0.3V, -8.5V
Input Voltage to GND, Any Input Pin ..... V- -0.3V, V+ +0.3V
Duration of Output Short Circuit to GND .... Indefinite
Continuous Total Power Dissipation (T_A = +70°C)
MAX270:
  Plastic DIP (derate 8mW/°C above +70°C) 640mW
  Wide SO (derate 10mW/°C above +70°C) 800mW
  CERDIP (derate 11.1mW/°C above +70°C) 889mW
MAX271:
  Plastic DIP (derate 8.7mW/°C above +70°C) 696mW
  Wide SO (derate 11.7mW/°C above +70°C) 941mW
  CERDIP (derate 12.5mW/°C above +70°C) 1000mW
Operating Temperature Ranges:
  MAX27_C_ _ .... 0°C to +70°C
  MAX27_E_ _ .... -40°C to +85°C
  MAX27_M_ _ .... -55°C to +125°C
Storage Temperature Range .... -65°C to +165°C
Lead Temperature (soldering, 10 sec) .... +300°C

Stresses beyond those listed under "Absolute Maximum Ratings" may cause permanent damage to the device. These are stress ratings only, and functional operation of the device at these or any other conditions beyond those indicated in the operational sections of the specifications is not implied. Exposure to absolute maximum rating conditions for extended periods may affect device reliability.

ELECTRICAL CHARACTERISTICS
(V+ = 5V, V- = -5V; T_A = +25°C, unless otherwise noted.)

PARAMETER	CONDITIONS		MIN	TYP	MAX	UNITS
FILTER CHARACTERISTICS						
Operating Frequency Range	(Note 1)				2	MHz
Programmed Cutoff Frequency (f_C) Range				1-25		kHz
Programmed Cutoff Frequency Error	f_C code = 53 (2.536kHz typ)			±2.9		%
	f_C code = 127 (25kHz typ)			±9.5		
Filter Gain	f_C code = 0 (1kHz typ), T_A = T_{MIN} to T_{MAX}	f_{IN} = 1kHz	-3.6		-2.4	dB
		f_{IN} = 8kHz			-33	
	f_C code = 127 (25kHz typ), T_A = T_{MIN} to T_{MAX}	f_{IN} = 25kHz	-6		-0.5	
		f_{IN} = 200kHz			-34	
Maximum Gain (Peaking)	f_C code = 0 (1kHz typ)				0.15	dB
	f_C code = 127 (25kHz typ)			0.15		
Wideband Noise	50Hz to 50kHz Bandwidth	f_C code = 0 (1kHz typ)		12		μV_{RMS}
		f_C code = 127 (25kHz typ)		38		
DC CHARACTERISTICS						
DC Output Signal Swing OUTA, OUTB, OP OUT (MAX270) OUTA, OUTB, T/H OUT (MAX271)	R_{LOAD} = 5kΩ, T_A = T_{MIN} to T_{MAX}		-3		3	V
Offset Voltage at Outputs OUTA, OUTB, OP OUT (MAX270) OUTA, OUTB (MAX271)			-2		2	mV
DC Input Leakage Current INA, INB (MAX270) INA, INB (MAX271)	T_A = T_{MIN} to T_{MAX}		-1		1	μA

Figure 3–7. Electrical characteristics for MAX270/271. (Maxim New Releases Data Book, 1992, pp. 6-62, 6-63) (*Figure continued on next page.*)

3.1.1 Programmable Capacitors

Each filter section contains two banks of programmable capacitors, controlled by an internal 7-bit memory, which set filter cutoff frequencies (fc) from 1-kHz to 25-kHz. The filters provide two program modes. In the μP mode, cutoff frequencies are programmed by writing 7-bit data to one of two memory addresses (one for each filter section). The pin-strap programming mode programs both filter sections simultaneously. With pin-strap, both memory latches are transparent (not addressable), and data pins D0–D6 can be pin strapped (hardwired) to set a common fc for both filter sections.

52 SIMPLIFIED DESIGN OF FILTER CIRCUITS

PARAMETER	CONDITIONS	MIN	TYP	MAX	UNITS
DYNAMIC FILTER CHARACTERISTICS - MAX270					
Total Harmonic Distortion (THD)	f_C code = 44 (2.01kHz typ), V_{IN} = 3.5V_{p-p} at 390.625Hz (Notes 2, 3)			-70	dB
Signal/(Noise + Distortion) (SINAD)			73		dB
Spurious-Free Dynamic Range (SFDR)		70			
UNCOMMITTED AMPLIFIER - MAX270					
Slew Rate			1.2		V/μs
Bandwidth			2		MHz
TRACK-AND-HOLD - MAX271					
Hold Settling Time	To 0.1% (Note 4)		500		ns
Acquisition Time	To 0.1% (Note 5)		1.8		μs
Hold Step			1		mV
Droop Rate	$T_A = T_{MIN}$ to T_{MAX}		30		μV/μs
Offset Voltage at T/H OUT	Includes filter offset	-6		6	mV
T/H OUT Disabled Output Leakage Current	$T_A = T_{MIN}$ to T_{MAX}, $\overline{T}/H$ = 0V (Track Mode)	-10		10	μA
Total Harmonic Distortion (THD)	f_C code = 44 (2.01kHz typ), V_{IN} = 3.5V_{p-p} at 390.625Hz, Sampling rate = 50kHz (Notes 2, 6, 7)			-70	dB
Spurious-Free Dynamic Range (SFDR)		70			
DIGITAL INPUTS					
Digital Input High Voltage	$T_A = T_{MIN}$ to T_{MAX} (Note 8)	2.4			V
Digital Input Low Voltage				0.8	V
Digital Input Current	$T_A = T_{MIN}$ to T_{MAX}, Digital input held at ±5V, includes MODE (MAX271) (Note 8)	-1		1	μA
POWER REQUIREMENTS					
Supply Voltage Range			±2.375 to ±8		V
Supply Current	$T_A = T_{MIN}$ to T_{MAX} (Note 9)			6.5	mA
Shutdown Supply Current	$T_A = T_{MIN}$ to T_{MAX} (Note 10)			15	μA
Power-Supply Rejection Ratio (PSRR) at 1kHz	f_C code = 0 (1kHz typ), V+ = 5VDC + 100mVp-p at 1kHz		30		dB

Note 1: All internal amplifiers limited to 2MHz bandwidth.
Note 2: Only filter A tested for these parameters.
Note 3: Spurious-Free Dynamic Range is the ratio of the fundamental to the largest of any harmonic or noise spur in dB.
Note 4: Includes T/H propagation delays. With 5kΩ, parallel 100pF load.
Note 5: ±2V input step settling 0.1% with 5kΩ parallel 100pF load.
Note 6: T/H pin toggled at sampling rate, 50% duty cycle.
Note 7: THD and SFDR specifications for T/H include contributions from filter.
Note 8: Digital pins include $\overline{SHDN}$, $\overline{WR}$, $\overline{CS}$, A0, D0-D6 (MAX270) and $\overline{SHDN}$, T/H A/$\overline{B}$, $\overline{WR}$, T/H EN, $\overline{CS}$, A0, A1, D0-D6, $\overline{T}/H$ (MAX271).
Note 9: Input of uncommitted op amp floating with a 5kΩ feedback resistor from input to output.
Note 10: $\overline{WR}$, $\overline{CS}$, A0, D0-D6 held at +5V; $\overline{SHDN}$ = 0V (MAX270). $\overline{WR}$, $\overline{CS}$, A0, A1, D0-D6, TH, T/H A/$\overline{B}$, T/H, MODE held at +5V; $\overline{SHDN}$ = 0V (MAX271).

Figure 3-7. Continued.

TIMING CHARACTERISTICS (Figure 2)
(V+ = 5V, V- = -5V; T_A = +25°C, unless otherwise noted.)

PARAMETER	SYMBOL	CONDITIONS	MIN	TYP	MAX	UNITS
$\overline{CS}$ to $\overline{WR}$ Setup	t_{ws}		0			ns
$\overline{CS}$ to $\overline{WR}$ Hold	t_{wh}		0			ns
$\overline{WR}$ Pulse Width	t_{wv}		100			ns
Address-Setup Time	t_{as}		30			ns
Address-Hold Time	t_{ah}		10			ns
Data-Setup Time	t_{ds}		30			ns
Data-Hold Time	t_{dh}		10			ns

Note 11: All input control signals specified with $t_r = t_f$ = 5ns (10% to 90% of +5V) and timed from a +1.6V voltage level.

Figure 3-8. Timing characteristics for MAX270/271. (Maxim New Releases Data Book, 1992, p. 6-64)

Continuous (Active) Lowpass Filters

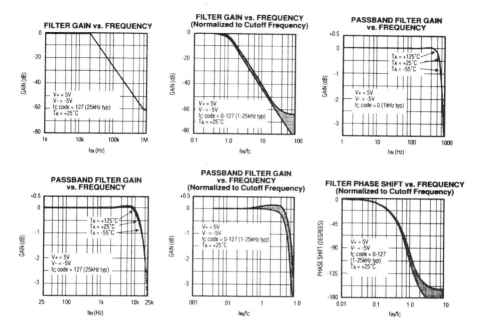

Figure 3-9. Operating characteristics for MAX270/271. (Maxim New Releases Data Book, 1992, pp. 6-64, 6-65) (*Figure continued on next page.*)

3.1.2 Filter Q and Passband

The filters are trimmed at the wafer level, setting Q for a maximum of 0.15 dB passband peaking for fc programmed at 1 kHz. Maximum passband peaking at other codes is typically less than 0.15 dB. Filter Q is not user programmable.

3.1.3 Op Amp and T/H

The MAX270 includes an uncommitted op amp (inverting input grounded). The MAX271 has an on-chip T/H that tracks and holds the output of either filter section (selectable). The help output is provided at T/H OUT (pin 1). T/H functions are controlled by writing control bits to internal registers (in μP mode) or by control pins directly (in pin-strap mode).

3.1.4 Quiescent Current and Shutdown

The ICs provide a low-quiescent-current shutdown mode controlled by the $\overline{SHDN}$ pin (4). A low at the $\overline{SHDN}$ pin turns off internal amplifiers and floats all outputs, reducing quiescent operating current to less than 15 μA. When the MAX271 is in the μP mode, shutdown mode is selected by writing control bits to memory (the $\overline{SHDN}$ pin is disabled).

54 SIMPLIFIED DESIGN OF FILTER CIRCUITS

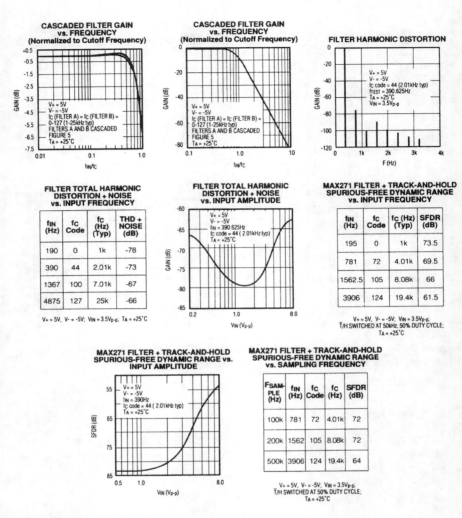

Figure 3-9. Continued.

3.2 Programming the Cutoff Frequency

The table of Fig. 3–13 shows how data pins D0–D6 provide for programming of 128 cutoff frequencies from 1 kHz to 25 kHz. Keep in mind that fc is the frequency of 3-dB attenuation in the filter response.

Continuous (Active) Lowpass Filters 55

PIN #	NAME	FUNCTION
1	OP OUT	Uncommitted Op-Amp Output
2	V+	Positive Supply Voltage
3	OUTA	Filter A Output
4	$\overline{\text{SHDN}}$	SHUTDOWN Control. Low level floats OUTA, OUTB, and OP OUT and places device into shutdown mode.
5	INA	Filter A Input
6	V-	Negative Supply Voltage
7	INB	Filter B Input
8	OUTB	Filter B Output
9	GND	Ground
10	$\overline{\text{WR}}$	WRITE Control Input. A low level writes data D0-D6 to program memory addressed by A0. High level latches data.
11	$\overline{\text{CS}}$	CHIP SELECT Input. Must be low for $\overline{\text{WR}}$ input to be recognized.
12	A0	Three-Level Address Input—logic high: addresses filter A; logic low: addresses filter B; connect to V-: pin-strap mode
13-19	D0-D6	7-Bit Data Inputs. Allows programming of 128 cutoff frequencies in a 1kHz to 25kHz range.
20	OP IN	Uncommitted Op-Amp Input

Note: All digital input levels are TTL and CMOS compatible, unless otherwise stated.

Figure 3-10.
Pin descriptions for MAX270. (Maxim New Releases Data Book, 1992, p. 6-66)

The equations for calculating fc from the programmed code are as follows:

$$fc = \frac{87.5}{87.5 - \text{CODE}} \times 1 \text{ kHz for codes 0-63 (fc = 1 kHz to 3.57 kHz)}$$

$$fc = \frac{262.5}{137.5 - \text{CODE}} \times 1 \text{ kHz for codes 64-127 (fc = 3.57 kHz to 25 kHz)}$$

where CODE is the data on pins D0–DC (0–127), and D6 is the most significant bit (MSB).

56 SIMPLIFIED DESIGN OF FILTER CIRCUITS

PIN #	NAME	FUNCTION, μP MODE (MODE = GND OR V-)	FUNCTION, PIN-STRAP MODE (MODE = V+)
1	T/H OUT	Track-and-Hold Output	
2	V+	Positive Supply Voltage	
3	OUTA	Filter A Signal Output	
4	$\overline{\text{SHDN}}$	X	$\overline{\text{SHUTDOWN}}$ Control. A low level floats outputs and places device into shutdown mode.
5	INA	Filter A Signal Input	
6	V-	Negative Supply Voltage	
7	INB	Filter B Signal Input	
8	MODE	Selects μP mode when tied to GND or V- and pin-strap mode when connected to V+.	
9	OUTB	Filter B Signal Output	
10	GND	Ground	
11	T/H A/$\overline{\text{B}}$	X	Track-and-Hold Input Control. A high/low level internally connects OUTA/OUTB to input of Track-and-Hold.
12	$\overline{\text{WR}}$	WRITE Control Input. A low level writes data D0-D6 to program memory addressed by A1, A0 (or performs function as described for address inputs). High level latches data.	X
13	T/H EN	X	Track-and-Hold Output Control. Low level floats T/H OUT. Connect pin high for normal operation.
14	$\overline{\text{CS}}$	CHIP SELECT Input. Must be low for $\overline{\text{WR}}$ input to be recognized.	X

15, 16	A1, A0	Address and μP Control Inputs.
		0, 0 Programs f_C, filter A.
		0, 1 Programs f_C, filter B.
		1, 0 Controls T/H functions:
		  D0 performs T/H EN pin function.
		  D1 performs T/H A/$\overline{\text{B}}$ pin function.
		1, 1 Controls device shutdown:
		  D0 performs $\overline{\text{SHDN}}$ pin function.
		Note: The $\overline{\text{WR}}$ pin must be strobed low to initiate a program/function (Figure 2).
			X

17-23	D0-D6	7-bit Data Inputs. Allows programming of 128 cutoff frequencies (also performs control functions as described above).	7-bit Data Inputs. Program memory latches are transparent in this mode. Connect pins high or low to program filters A and B simultaneously to the same f_C.
24	T/H	Track-and-Hold Control. Low level causes T/H OUT to track selected filter output. Filter output level held at T/H OUT synchronous with T/H rising transition.	

X = Pin has no function in this mode.
Note: All digital input levels are TTL and CMOS compatible, unless otherwise stated.

Figure 3-11. Pin descriptions for MAX271. (Maxim New Releases Data Book, 1992, p. 6-67)

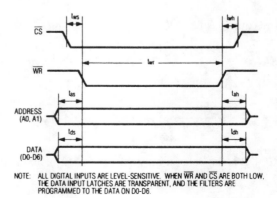

Figure 3-12.
Timing diagram for MAX270/271. (Maxim New Releases Data Book, 1992, p. 6-71)

Continuous (Active) Lowpass Filters **57**

PROGRAMMED CODE	fc (kHz)	PROGRAMMED CODE	fc (kHz)	PROGRAMMED CODE	fc (kHz)	PROGRAMMED CODE	fc (kHz)
0	1.000	32	1.576	64	3.571	96	6.325
1	1.011	33	1.605	65	3.620	97	6.481
2	1.023	34	1.635	66	3.671	98	6.645
3	1.035	35	1.666	67	3.723	99	6.818
4	1.047	36	1.699	68	3.777	100	7.008
5	1.060	37	1.732	69	3.832	101	7.191
6	1.073	38	1.767	70	3.888	102	7.394
7	1.087	39	1.804	71	3.947	103	7.608
8	1.100	40	1.842	72	4.007	104	7.835
9	1.114	41	1.881	73	4.069	105	8.076
10	1.129	42	1.923	74	4.133	106	8.333
11	1.143	43	1.966	75	4.200	107	8.606
12	1.158	44	2.011	76	4.268	108	8.898
13	1.174	45	2.058	77	4.338	109	9.210
14	1.190	46	2.108	78	4.411	110	9.545
15	1.206	47	2.160	79	4.487	111	9.905
16	1.223	48	2.215	80	4.565	112	10.294
17	1.241	49	2.272	81	4.646	113	10.714
18	1.259	50	2.333	82	4.729	114	11.170
19	1.277	51	2.397	83	4.816	115	11.666
20	1.296	52	2.464	84	4.906	116	12.209
21	1.315	53	2.536	85	5.000	117	12.804
22	1.335	54	2.611	86	5.097	118	13.461
23	1.356	55	2.692	87	5.198	119	14.189
24	1.378	56	2.777	88	5.303	120	15.000
25	1.400	57	2.868	89	5.412	121	15.909
26	1.422	58	2.966	90	5.526	122	16.935
27	1.446	59	3.070	91	5.645	123	18.103
28	1.470	60	3.181	92	5.769	124	19.444
29	1.495	61	3.301	93	5.898	125	21.000
30	1.521	62	3.431	94	6.034	126	22.826
31	1.548	63	3.571	95	6.176	127	25.000

Programmed code is the data on pins D0-D6 (0-127). D6 is the MSB.

Figure 3-13. Programmed cutoff frequency codes. (Maxim New Releases Data Book, 1992, p. 6-70)

For example, assume that a cutoff frequency fc of 2 kHz is required. In that case, use program code 43 (or 44) shown in Fig. 3-13. This can be confirmed using the equation:

$$fc = \frac{87.5}{87.5 - 43} \times 1 \text{ kHz} = \frac{87.5}{44.5} \times 1 \text{ kHZ} = 1.966 \text{ kHZ}$$

Obviously, the table is much easier to use. However, the author will permit you to use the equation (and a calculator) if you prefer.

The final step is to convert the program code of 43 into a 7-bit digital code D0–D6, with D6 as the MSB, or 0101011. If you are not familiar with the digital numbering systems, read Lenk's *Digital Handbook,* McGraw-Hill, 1993, written by this author. The code 0101011 can be entered into the IC by means of pin-strap connections, or from a μP, at pins 17–23.

3.2.1 Program Accuracy

The actual cutoff frequencies are subject to some error for each programmable code. The highest accuracy occurs at CODE = 0 where filters are trimmed for a 1-kHz cutoff frequency. At higher codes, the CODE versus fc errors increase. However, the frequency error at CODE = 127 (highest code) remains typically within ±9.5 percent. This means that the actual filter cutoff frequency, when programmed to CODE = 127, falls between 22.63 kHz and 27.38 kHz. It also means that you must use all calculations (and the values that result from such calculations) as a starting point for design (as the first trial values). This caution applies to all calculations (and values) in this book!

3.3 MAX270 Control Interface

Figure 3–14 shows the control-interface logic for the MAX270. (Interface timing for both ICs in the μP mode is shown in Fig. 3–12.) The AO pin (12) is a three-level input that selects the memory addresses for updating cutoff-frequency data in the μP mode. Connecting AO to the negative supply selects pin-strap mode (where there are no timing requirements). With pin-strap, the internal memory latches are disabled, permitting filters A and B to be programmed directly by the fc data strapped on pins D0–D6. The pin-strap mode disables $\overline{CS}$ (chip select) and $\overline{WR}$ (write) controls. Both filters A and B are programmed to the same fc.

A low level on the $\overline{SHDN}$ pin (4) shuts down all amplifiers and floats OUTA, OUTB, and OP OUT. Current consumption drops to less than 15 μA in the shutdown mode.

3.4 MAX271 Control Interface

Figure 3–15 shows the μP-mode control-interface logic for the MAX271. Connecting the MODE pin (8) to GND or V- selects the μP mode. With the μP mode, an addressable program memory controls filter cutoff frequency programming, and all T/H functions, except $\overline{T/H}$. In the μP mode, $\overline{SHDN}$, T/H A/B, and T/H EN pins are disabled. $\overline{T/H}$ remains enabled and performs the T/H tracking/holding function.

Figure 3–14.
Control interface logic for MAX270. (Maxim New Releases Data Book, 1992, p. 6-71)

A0	SELECTS
Logic Low	Filter B
Logic High	Filter A

Continuous (Active) Lowpass Filters

A1	A0	D6	D5	D4	D3	D2	D1	D0	FUNCTION
0	0	\multicolumn{7}{c}{7-bit fc data}								Selects filter A
0	1	\multicolumn{7}{c}{7-bit fc data}								Selects filter B
1	0	X	X	X	X	X	X	0	T/H OUT disabled
1	0	X	X	X	X	X	X	1	T/H OUT enabled
1	0	X	X	X	X	X	0	X	Selects OUTB as input to T/H
1	0	X	X	X	X	X	1	X	Selects OUTA as input to T/H
1	1	X	X	X	X	X	X	0	Filter shutdown mode. All outputs floated, 15μA max supply current
1	1	X	X	X	X	X	X	1	Removes filter from shutdown mode

X = Don't care

Figure 3-15. Control interface logic for MAX271 in μP mode. (Maxim New Releases Data Book, 1992, p. 6-71)

Tying the MODE pin (8) to V+ selects the pin-strap mode. With pin-strap, both memory latches are transparent, and data bits on D0–D6 control the fc of filters A and B directly (filters A and B are programmed to the same fc). A0, A1, $\overline{CS}$, and $\overline{WR}$ are disabled during the pin-strap mode.

3.5 Digital Threshold Levels

Figure 3-16 shows the logic (or digital) threshold voltage levels for various supply voltages. Typically, the logic-voltage thresholds are a fraction of the V+ supply. All digital inputs are TTL and CMOS compatible, unless otherwise stated. Inputs are CMOS gates with less than 1-μA leakage current and 8-pF capacitance loading.

3.6 Filter Performance

All of the internal amplifiers, output stages for filter sections, uncommitted op amp (MAX270), and T/H (MAX271) are identical. The outputs are designed to drive

V+ (V)	LOGIC THRESHOLD VOLTAGE (V)
8	+2.4
7	+2.3
6	+2.0
5	+1.75
4	+1.5
2.5	+1.0

Figure 3-16. Logic threshold voltage levels. (Maxim New Releases Data Book, 1992, p. 6-72)

NOTE: For +5V single-supply operation, where incoming logic signals are referenced to V-, typical logic thresholds are +3.5V. Therefore, a CMOS (rail-to-rail) logic interface is recommended.

5-k loads in parallel with a maximum capacitance of 100 pF. At higher load levels, the output swing becomes asymmetric. All outputs can be short-circuited to GND for an indefinite duration. The operating frequency range is limited to about 2 mHz by the bandwidth of the internal amplifiers.

3.6.1 MAX270/271 Noise

Wideband filter noise over a 50-kHz bandwidth is 12 μV(rms) and 38 μV(rms) per section for fc programmed to 1 kHz and 25 kHz, respectively. This results in a dynamic range of over 96 dB.

3.6.2 Filter Input Impedance

At DC, the input impedance at INA and INB is equal to the DC input impedance of the amplifier (about 5 M). At higher frequencies, internal capacitors contribute to an effective input impedance that might fall as low as 100 k at 25 kHz.

3.7 MAX271 Track-and-Hold

The MAX271 T/H is functionally equivalent to a switched 200-pF capacitor buffered by a unity-gain amplifier (Figs. 3–2, 3–3). When the $\overline{\text{T/H}}$ pin is driven low, the output of filter A or filter B (whichever is selected via the control interface) internally connects to the amplifier, and T/H OUT follows the filter output.

The offset at T/H OUT (±6 mV maximum) is the combined offset of the filter amplifier and the T/H buffer. When $\overline{\text{T/H}}$ is pulled high, the switch disconnects the filter signal from the T/H. The T/H capacitor holds the stored charge, and that voltage is buffered at T/H OUT.

3.7.1 Multiplexed Operation for MAX271

Figure 3–17 shows how the MAX271 can be connected for multiplex operation. For multiplex, a low level at T/H EN floats T/H OUT. T/H A/$\overline{\text{B}}$ selects between OUTA and OUTB for the T/H input. In the μP mode, the T/H EN and T/H OUT functions are controlled by writing control bits to program memory, with the T/H EN and T/H OUT pins disabled.

3.8 Power-Supply Configurations

Figure 3–18 shows the recommended power-supply connections. As always, the supplies must be properly bypassed. Best performance is achieved if V+ and V– are bypassed to ground with 4.7-μF electrolytic (tantalum is preferred) and 0.1-μF ceramic capacitors in parallel. These should be as close as possible to the chip supply pins.

Continuous (Active) Lowpass Filters **61**

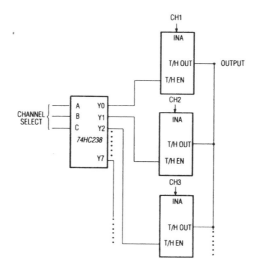

Figure 3–17.
MAX271 connected for multiplex operation. (Maxim New Releases Data Book, 1992, p. 6-73)

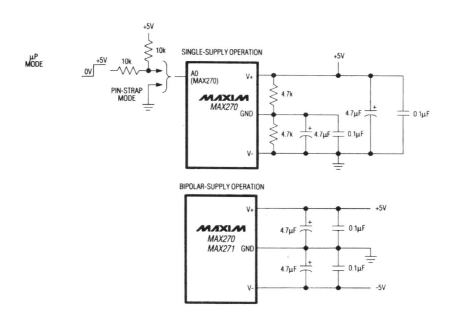

Figure 3–18. Recommended power-supply connections for MAX270/271. (Maxim New Releases Data Book, 1992, p. 6-73)

62 SIMPLIFIED DESIGN OF FILTER CIRCUITS

Single supplies in the range of 4.75 V to 16 V can be used. Digital logic can be referenced to V– (system ground), but will not maintain TTL compatibility. CMOS (rail-to-rail) logic is recommended. For µP-mode operation with a single supply, the MAX270 A0 pin must be configured with a voltage divider as shown in Fig. 3–18. The lowest quiescent current in the shutdown mode is achieved when A0 is either at V+ or V–.

3.9 Programming without a Microprocessor

Figure 3–19 shows how filter sections A and B can be programmed to different cutoff frequencies without a microprocessor. The MAX690 µP-supervisory IC provides the proper programming sequence when the circuit is first powered up. This is done by controlling the 74HC373 data buffer and the MAX270 addressing pin to load independent fc data for filters A and B. If you are not familiar with the operation of µP-supervisory ICs, read the author's *Simplified Design Microprocessor Supervisory Circuits,* Butterworth-Heinemann, 1997.

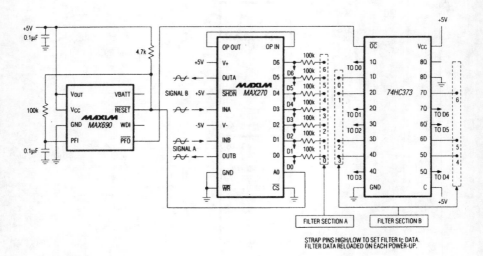

Figure 3–19. Programming without a microprocessor. (Maxim New Releases Data Book, 1992, p. 6-74)

3.10 Typical Application (Cascading)

Figure 3–20 shows how both sections of the MAX270 can be cascaded to provide an 80-dB/decade rolloff (4th-order). The input signal is applied at the input to filter section A, producing a 40-dB/decade rolloff. The output from section A is applied to section B, resulting in an 80-dB/decade rolloff at the output of section B. The pin-strap programming pins are all connected to ground (D0–D6 = digital 0000000), resulting in a program code of 0 (Fig. 3–13) and an fc of 1 kHz.

The $\overline{\text{SHDN}}$ pin is connected to V+ to keep the IC operating. The A0 pin is connected to V– to select the pin-strap mode. Because dual supplies are available, use the bipolar-supply bypassing configuration shown in Fig. 3–18. The $\overline{\text{CS}}$ and $\overline{\text{WR}}$ pins are connected to GND to disable the chip-select and write functions. OP IN is also connected to ground, thus disabling the on-chip uncommitted op amp.

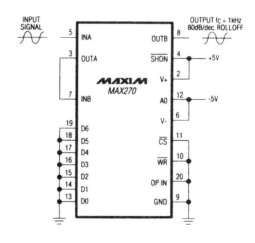

Figure 3–20.
Cascading sections to provide an 80-dB/decade rolloff. (Maxim New Releases Data Book, 1992, p. 6-74)

CHAPTER **4**

Zero DC-Error Lowpass Filters

This chapter is devoted to simplified design with typical switched-capacitor lowpass IC filters (the Maxim MAX280 and Linear Technology LTC1062). The MAX280 is an enhanced version of the LTC1062. The enhancements (according the Maxim) include tighter specifications on the internal clock-oscillator frequency and the buffer-amplifier offset voltage. Both ICs require few external components and can be designed to provide a wide range of cutoff frequencies using simple equations.

The MAX280/LTC1062 is a 5th-order lowpass filter with no DC error. The ICs use an external resistor and capacitor to isolate the IC from the DC signal path, thus providing DC accuracy. The resistor and capacitor, along with the on-chip 4th-order switched-capacitor filter, form a 5th-order lowpass filter. Two ICs can be cascaded to form a 10th-order lowpass function. The filter cutoff frequency is set by an internal clock, which can be externally driven. The clock-to-cutoff frequency ratio of 100:1 allows clock ripple to be easily removed.

4.1 Basic Circuit Functions for MAX280/LTC1062

Figure 4–1 shows the internal circuits in block form. Figures 4–2 and 4–3 show a typical operating circuit and pin configuration, respectively. Figures 4–4, 4–5, and 4–6 show the electrical, pin descriptions and operating characteristics, respectively.

As shown in Fig. 4–1, the output voltage is sensed through an internal buffer, then applied to an internal switched-capacitor network that drives the bottom plate of an external capacitor to form a 5th-order lowpass filter. The input and output appear across an external resistor. The IC part of the overall filter handles only the AC path of the signal. The DC offsets of the buffer and the switched-capacitor network are blocked by this capacitor and do not appear at the zero-offset output pin.

As discussed in Section 1.11.8, switched-capacitor filters are subject to aliasing problems. The external resistor and capacitor shown in Fig. 4–1 automatically pro-

65

66 SIMPLIFIED DESIGN OF FILTER CIRCUITS

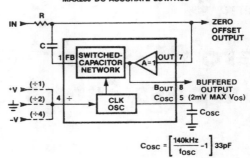

Figure 4-1.
Internal circuits of MAX280. (Maxim New Releases Data Book, 1992, p. 6-83)

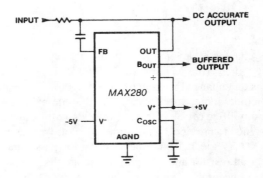

Figure 4-2.
Typical operating circuit of MAX280. (Maxim New Releases Data Book, 1992, p. 6-79)

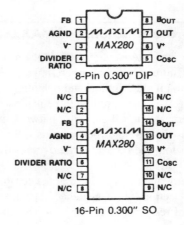

Figure 4-3.
Pin configurations of MAX280. (Maxim New Releases Data Book, 1992, p. 6-79)

ABSOLUTE MAXIMUM RATINGS

Total Supply Voltage (V⁺ to V⁻) 18V
Input Voltage at Any Pin ...... V⁻ -0.3V ≤ V$_{IN}$ ≤ V⁺ +0.3V
Operating Temperature
  MAX280CXX/LTC1062C .... -0°C to +70°C
  MAX280EXX .... -40°C to +85°C
  MAX280MXX/LTC1062M .... -55°C to +125°C
Storage Temperature .... -65°C to +160°C
Lead Temperature Range (Soldering, 10 sec) ..... +300°C
Power Dissipation
  Plastic DIP (derate 6.25mW/°C above 70°C) 500mW
  CERDIP (derate 8.00mW/°C above 70°C) 640mW
  SO (derate 9.52mW/°C above 70°C) 762mW

Stresses above those listed under "Absolute Maximum Ratings" may cause permanent damage to the device. These are stress ratings only and functional operation of the device at these or any other conditions above those indicated in the operational sections of the specification is not implied. Exposure to absolute maximum rating conditions for extended periods may affect device reliability.

ELECTRICAL CHARACTERISTICS

(V⁺ = +5V, V⁻ = -5V, T$_A$ = 25°C, unless otherwise specified, AC output measured at pin 7, Figure 1.)

PARAMETER	CONDITIONS	MIN	TYP	MAX	UNITS

Operating Supply Voltage
  Dual Supply
  Single Supply
		±2.375
		4.75
				±8.0
				16.0
					V

Power Supply Current	C$_{OSC}$ (Pin 5 to V⁻) = 100pF T$_A$ = 25°C T$_A$ = T$_{MIN}$ to T$_{MAX}$		5.0 5.0	7.0 10.0	mA
Input Frequency Range			0-20		kHz
Filter Gain at f$_{IN}$ = 0 f$_{IN}$ = 0.5f$_C$ (Note 1) f$_{IN}$ = f$_C$ f$_{IN}$ = 2f$_C$ f$_{IN}$ = 4f$_C$	f$_{CLK}$ = 100kHz, Pin 4 at V⁺ C = 0.01µF, R = 25.78kΩ T$_A$ = T$_{MIN}$ to T$_{MAX}$ T$_A$ = T$_{MIN}$ to T$_{MAX}$ T$_A$ = T$_{MIN}$ to T$_{MAX}$	 -2 -28 -54	0 -0.02 -3 -30 -60	 -0.3	dB
Clock to Cutoff Frequency Ratio f$_{CLK}$/f$_C$	f$_{CLK}$ = 100kHz, Pin 4 at V⁺ C = 0.01µF, R = 25.78kΩ		100 ± 1		
Filter Gain at f$_{IN}$ = 16kHz	f$_{CLK}$ = 400kHz, Pin 4 at V⁺ C = 0.01µF, R = 6.5kΩ T$_A$ = T$_{MIN}$ to T$_{MAX}$	-48	-52		dB
f$_{CLK}$/f$_C$ Tempco	Same as above		10		ppm/°C
Filter Output (Pin 7) DC Swing	Pin 7 buffered with an ext op amp T$_A$ = T$_{MIN}$ to T$_{MAX}$	±3.5	±3.8		V
Clock Feedthrough			10		mV$_{pp}$
INTERNAL BUFFER					
Bias Current	T$_A$ = 25°C T$_A$ = T$_{MIN}$ to T$_{MAX}$		2 170	50 1000	pA
Offset Voltage	MAX280 LTC1062		0.2 2	2 20	mV
Voltage Swing	R1 = 20kΩ; T$_A$ = T$_{MIN}$ to T$_{MAX}$	±3.5	±3.8		V
Short Circuit Current Source/Sink			30/2		mA
CLOCK (NOTE 2)					
Internal Oscillator Frequency	C$_{OSC}$ (Pin 5 to V⁻) = 100pF	MAX280 LTC1062	31 25	35 35	39 50	kHz
	T$_A$ = T$_{MIN}$ to T$_{MAX}$ C$_{OSC}$ (Pin 5 to V⁻) = 100pF	MAX280 LTC1062	29 15	35 35	43 65	
Max Clock Frequency			4		MHz
C$_{OSC}$ Input Sink/Source Current	T$_A$ = T$_{MIN}$ to T$_{MAX}$		25	80	µA

Note 1: f$_C$ is the frequency where the gain is -3dB with respect to the input signal.
Note 2: The external or driven clock frequency is divided by either 1, 2, or 4 depending upon the voltage at pin 4. When pin 4 = V⁺, f$_{CLK}$/f$_C$ = 100; when pin 4 = GND, f$_{CLK}$/f$_C$ = 200; pin 4 = V⁻, f$_{CLK}$/f$_C$ = 400.

Figure 4-4. Electrical characteristics of MAX280. (Maxim New Releases Data Book, 1992, p. 6-80)

Pin Description

PIN #	NAME	FUNCTION
1	FB	External capacitor couples to the chip through this pin.
2	AGND	Ground. Connect to system ground for dual supply operation or mid-supply for single operation. This pin should be well bypassed using a large capacitor for single supply operation.
3	V⁻	Negative supply voltage
4	DIVIDER RATIO	The oscillator frequency is divided by either 1, 2, or 4 depending upon the voltage on this pin. This in turn gives a clock to cutoff frequency ratio when tied to V⁺ of 100:1; when tied to GND of 200:1; and when tied to V⁻ of 400:1.

PIN #	NAME	FUNCTION
5	C_{OSC}	Clock input pin for external clock applications. For internal clock operation connect an external capacitor between this pin and V⁻.
6	V⁺	Positive supply voltage
7	OUT	Input to on-chip buffer amplifier
8	B_{OUT}	Output of buffer amplifier

Figure 4–5. Pin descriptions of MAX280. (Maxim New Releases Data Book, 1992, p. 6-81)

vide the required anti-aliasing. Also, low-frequency noise in the IC is attenuated by the external capacitor because any noise at the FB pin goes through a highpass path to the filter output.

The filter output is unbuffered. The output signal can be buffered by the on-chip buffer or by a high-accuracy op amp (such as a chopper-stabilized op amp) to get a buffered DC-accurate signal. The on-chip buffer has an offset voltage of 2 mV for the MAX280 and 20 mV for the LTC1062. The offset voltages for both devices have a typical TC of 1-μV/°C.

4.2 Using the Clock Divider Ratio

The DIVIDER RATIO pin (4) sets the ratio between the internal fCLK (supplied to the IC) and fOSC (the output at the DIVIDER RATIO pin). Connect the DIVIDER RATIO pin to V+ for a 1:1, to GND for a 1:2, or to V– for a 1:4 fCLK/fOSC ratio.

4.3 Using the Internal Oscillator

The internal 140-kHz (nominal) oscillator frequency can be modified by connecting an external capacitor in parallel with the on-chip 33-pF capacitor. Connect the external capacitor to GND if non-polarized, or to V– if the capacitor is polarized.

The clock frequency can be calculated by

$$fOSC = 140 \text{ kHz } (33pF/33pF + Cosc)$$

Zero DC-Error Lowpass Filters 69

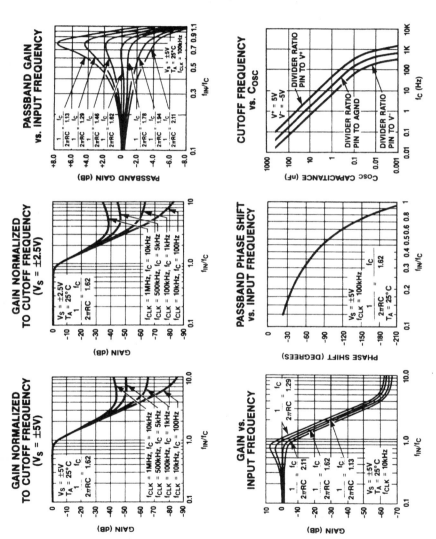

Figure 4-6. Typical operating characteristics of MAX280. (Maxim New Releases Data Book, 1992, pp. 6-81, 6-82) *(Figure continued on next page.)*

70 SIMPLIFIED DESIGN OF FILTER CIRCUITS

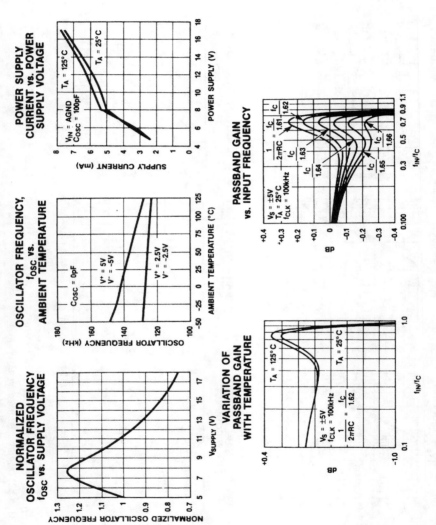

Figure 4-6. Continued.

Because of process tolerances, fOSC can vary by ±62.5 percent in the LTC1062. In the MAX280, on-chip trimming reduces the fOSC tolerance to ±19.5 percent.

The oscillator frequency can be adjusted by adding a series potentiometer between the capacitor and the Cosc pin as shown in Fig. 4–7. When an external pot is used, the new oscillator frequency is always higher than the one calculated by the fOSC equation. Using the values shown in Fig. 4–7, the fOSC can be adjusted from about 500 kHz to 1.5 kHz. The oscillation frequency can be measured directly at the Cosc pin using a low-capacitance probe.

4.4 Using an External Clock

The internal switched-capacitor filter requires a clock 100 times higher than the desired cutoff frequency. In an external clock is used, the input on the Cosc pin must swing close to the power rails (V+, V–). Although standard 74HC00-series CMOS gates do not guarantee CMOS levels with the source and sink currents of the Cosc pin, such gates will (in reality) drive the Cos pin.

CMOS gates conforming to standard B-series output drive have the appropriate voltage levels and current to simultaneously drive several chips. The typical trip levels of the internal Schmitt-trigger-sensing Cosc pin are as shown in Fig. 4–8.

4.5 Choosing External Resistor and Capacitor Values

The Passband Gain vs. Input Frequency graph of Fig. 4–6 can be used for selecting the values of external resistor R and capacitor C (Fig. 4–1) to produce the desired response and cutoff frequency. Both R and C affect the cutoff frequency fc (–3 dB point) as does the clock frequency. For a max-flat (typical Butterworth) response, the clock should be 100 times the desired fc. As an example, for a 10 Hz cutoff frequency, the clock should be 1 kHz.

With the clock frequency established, the next step is to divide the desired fc by the factor (1.61, 1.62, etc.) shown in the graph that corresponds to the desired response. (The manufacturer recommends that the fc be divided by 1.62 for Butterworth.) Then choose R and C to produce the same frequency, using this relationship: $fc/1.62 = 1/(2\pi RC) = 1/(6.28RC)$. To simplify the calculations, let C be some standard value such as 0.01 µF, 0.1 µF, 1 µF, and so on, for a realistic value of R.

Figure 4–7.
External oscillator trim circuit. (Maxim New Releases Data Book, 1992, p. 6-83)

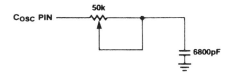

Figure 4-8.
Trip levels of internal
Schmitt-trigger-sensing
Cosc pin. (Maxim New
Releases Data Book,
1992, p. 6-83)

POWER SUPPLY		TRIP LEVEL	
V^+ = +2.5V	V^- = -2.5V	V_{IH} = 0.9V	V_{IL} = -1.15V
V^+ = +5.0V	V^- = -5.0V	V_{IH} = 1.4V	V_{IL} = -2.1V
V^+ = +6.0V	V^- = -6.0V	V_{IH} = 1.7V	V_{IL} = -2.5V
V^+ = +5.0V	V^- = 0V	V_{IH} = 3.4V	V_{IL} = 1.35V
V^+ = +10V	V^- = 0V	V_{IH} = 6.4V	V_{IL} = 2.9V
V^+ = +15V	V^- = 0V	V_{IH} = 9.5V	V_{IL} = 4.1V

To find the value of R, first divide fc by the desired response factor (1.62 in our case). With an fc of 10 Hz, the result is 10/1.62 = 6.17. Then divide 1 by 6.17 to find the desired product of 6.28RC, with a result of 1/6.17 = 0.162. Assume a value of 1 µF for C, then divide 0.162 by 6.28^{-6} for an R of 25.8 k.

The manufacturer recommends that R should be in the 20-k range. However, this can result in some large values of C, depending on the desired fc. For example, you might not be able to find an actual temperature-stable capacitor (such as an NPO ceramic) with a value of 1 µF (required for our 10-Hz fc). If you choose a 0.1 µF (which should be available with ±20-ppm temperature coefficients), the value of R must be raised to 258 k. As an alternate, Mylar, polystyrene, and polypropylene capacitors should provide acceptable performance.

The minimum value of R depends on the maximum input signal and the current-sinking capability of the FB pin (typically 1 mA). So, for a 1-V peak-to-peak signal, the minimum value of R is 1 k.

Regardless of what values are involved, R and C must be chosen carefully for the max-flat Butterworth response. (Design is easier if a Chebyshev response is acceptable.) This is because response peaking becomes severe when the R and C cutoff frequency approaches the fc of the on-chip 4th-order filter. This is shown in the Passband Gain vs. Input Frequency graph of Fig. 4–6. However, the attenuation slope is virtually unaffected by R and C because slope is set by internal circuits. This can be seen in the Gain vs. Input Frequency graph of Fig. 4–6.

4.6 Input Voltage Range for MAX280

Every node of filter typically swings with 1 V of both supplies. With the correct external resistor and capacitor values, the amplitude response of all the internal and external nodes should not exceed a gain of 0 dB, with the exception of the FB pin. Figure 4–9 shows the amplitude response of the FB pin. Note that there is some peaking.

For an input frequency around 0.8fc, the gain is 1.7V/V and with ±5-V supplies, the peak-to-peak input voltage should not exceed 4.7 V. If the input voltage goes beyond 4.7 V, clipping and distortion of the output waveform might occur. However, the filter will not be damaged. The absolute maximum input voltage to any pin should not exceed the power supplies.

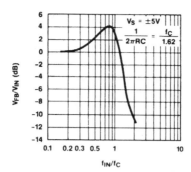

Figure 4–9.
Amplitude response of FB pin. (Maxim New Releases Data Book, 1992, p. 6-84)

4.7 Internal Buffer

The internal output buffer of the FB pin, and the OUT pin, is part of the AC signal path. As a result, capacitive loading greater than 30 pF might cause gain errors in the passband around the cutoff frequency. The internal buffer can also be used as the filter output. However, there will be a few millivolts of output offset.

4.8 Filter Attenuation

The normal rolloff is 30 dB/octave. When the clock rate is increased (with a corresponding increase in fc), the maximum attenuation of the filter decreases as shown in the typical operating characteristics of Fig. 4–5. This decrease is caused by rolloff at higher frequencies of the loop gains of the various internal feedback paths, and is not because of any increase in noise floor.

4.9 Filter Noise

The filter wideband noise is typically 90μVrms, with ±5-V supplies, and typically 80μVrms for ±2.5-V supplies. This noise value is nearly independent of fc. The noise spectral density, unlike active filters, is nearly zero for frequencies below 0.1 fc. About two-thirds of the entire wideband noise is in the band from DC to fc.

4.10 Transient Response for MAX280

Figure 4–10 shows the step response of the filter using the response characteristics of fc/1.62. This response approximates that of a 5th-order max-flat (Butterworth) filter. However, there is some ringing in response to the 1-V input pulse (or step). The ringing can be reduced using a response characteristic of fc/2 as shown in

74 SIMPLIFIED DESIGN OF FILTER CIRCUITS

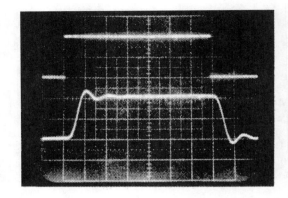

Figure 4–10.
Step response of Butterworth approximation.
(Maxim New Releases Data Book, 1992, p. 6-84)

$$\frac{1}{2\pi RC} = \frac{f_C}{1.62}, (1mS/div., 0.5V/div.)$$

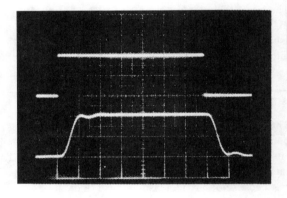

Figure 4–11.
Step response of Bessel approximation. (Maxim New Releases Data Book, 1992, p. 6-84)

$$\frac{1}{2\pi RC} = \frac{f_C}{2}, (1mS/div., 0.5V/div.)$$

Fig. 4–11. This is essentially a Bessel response. Chapter 5 discusses Bessel filters in greater detail.

4.11 Anti-Aliasing

The internal 4th-order filter is a sampled device (switched capacitor), and as such will alias unless preceded by a band-limited signal, or a continuous non-sampled filter. Fortunately, the external resistor R and capacitor C used to form the 5th-order automatically provides this function. Typical attenuation is greater than 35 dB at the Nyquist frequency.

4.12 LTC1062 Characteristics

Figure 4–12 shows the architecture and basic connections for the LTC1062. (Compare this to Fig. 4–1.) Again, the clock frequency should be 100 times fc, and the values of R and C set fc at the −3-dB point. Figure 4–13 shows the passband response for values of 1/6.28RC near fc/1.62. Figures 4–14 and 4–15 show additional response characteristics, but with a wider range of values.

If you study Figs. 4–13 through 4–15, you will see that even a slight change in component values can produce a drastic change in response characteristics. This is true for all electronic filters! As a result, use any combination of values obtained by any design means (equations, graphs, tables, even filter-design software) as *first trial values*. Then change the values (after testing) as needed to get the desired results.

Figure 4–16 shows the LTC1062 operated with the internal clock. If Cosc is 8500 pF and a 50-k pot is connected to pin 5 as shown, the clock frequency can be adjusted from about 500 Hz to 3.3 kHz (providing for an fc of 5 to 33 Hz).

4.13 Simplified Design Approaches (MAX280/LTC1062)

This section describes a variety of switched-capacitor circuits using the MAX280 and/or LTC1062. All of the design information discussed in this chapter applies to the examples in this section. However, each circuit has special design requirements, which are discussed in detail. The circuits can be used immediately the way they are or, by altering component values, as a basis for simplified design of similar filters.

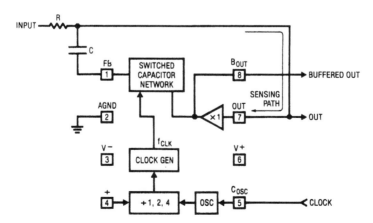

Figure 4–12. Architecture and basic connections for LTC1062. (Linear Technology, Application Note 20, p. 1)

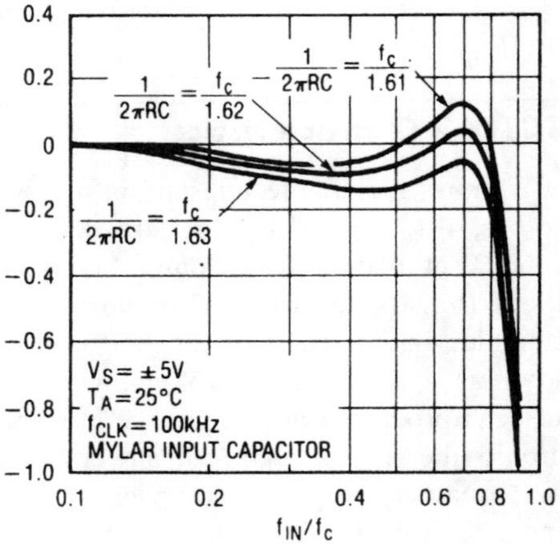

Figure 4–13.
Passband response for f_{IN}/f_c 0.1 to 1.0. (Linear Technology, Application Note 20, p. 3)

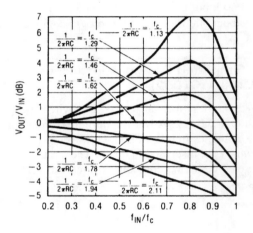

Figure 4–14.
Passband response for f_{IN}/f_c 0.2 to 1. (Linear Technology, Application Note 20, p. 3)

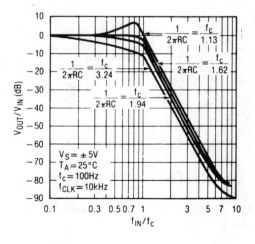

Figure 4–15.
Passband response for f_{IN}/f_c 0.1 to 10. (Linear Technology, Application Note 20, p. 4)

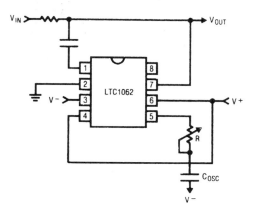

Figure 4-16.
LTC1062 operated with internal clock. (Linear Technology, Application Note 20, p. 2)

4.13.1 Single-Supply 5th-Order Lowpass Filter (MAX280)

Figure 4–17 shows a MAX280 connected to provide lowpass cutoff at 10 Hz. Other cutoff frequencies can be selected by altering the values of R and C. This circuit is similar to that described in Section 4.5, but with the MAX280 operated from a single +5-V supply.

Note that both pins 4 and 6 must be connected directly to +5 V. Pin 2 is fixed at +2.5 V by the two 25-k resistors, and the output at pin 7 is returned to pin 2 through a resistor. This feedback resistor R′ is 12 times the value of R, or 352.8 k.

With pin 4 connected to +5 V, the divider ratio is at 100:1, so an external clock of 1 kHz is required for a cutoff of 10 Hz. Note that the manufacturer recommends a response factor of fc/1.84 for single-supply operation. To find the value of R, divide fc (10) by 1.84 for a result of 10/1.84 = 5.434. Then divide 1 by 5.434 to find the de-

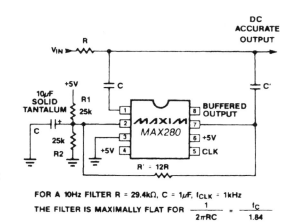

Figure 4-17.
MAX280 connected to provide lowpass cutoff at 10 Hz. (Maxim New Releases Data Book, 1992, p. 6-85)

78   SIMPLIFIED DESIGN OF FILTER CIRCUITS

sired product of 6.28RC, with a result of 1/5.434 = 0.184. Assume a value of 1 µF for C, then divide 0.184 by 6.28^{-6} for an R of 29.3 k or 29.4 k.

4.13.2   7th-Order Sallen/Key Lowpass Filter (MAX280)

Figure 4–18 shows a MAX280 and MAX430 amplifier connected to provide lowpass cutoff at 100 Hz. Other cutoff frequencies can be selected by altering the values of R and C connected to pin 1 of the MAX280. This circuit requires dual supplies, with pins 4 and 6 connected to +5 V and pin 3 connected to –5 V.

With pin 4 connected to +5 V, the divider ratio is at 100:1, so an external clock of 10 kHz is required for a cutoff of 100 Hz. Using the standard (Butterworth) response factor of fc/1.62, the result is 100/1.62 = 61.73. Then divide 1 by 61.73 to find the desired product of 6.28RC, with a result of 1/61.73 = 0.0162. Assume a value of 1 µF for C, then divide 0.0162 by 6.28^{-6} for an R of 2.6 k.

The output gain of the complete filter circuit is set by the values of R1–R4 and C1–C4 as shown by the table in Fig. 4–18. When no gain (unity gain) is required, simply omit R3 and R4, and connect the output of the MAX430 directly to the inverting input.

4.13.3   DC-Accurate Cascaded Lowpass Filter (LTC1062)

Figure 4–19 shows two LTC1062 filters where the second input is taken directly from the DC-accurate output of the first filter. The rolloff should be 60dB/octave, with 0-dB DC gain. The passband error is a typical 0.5 dB. The recommended

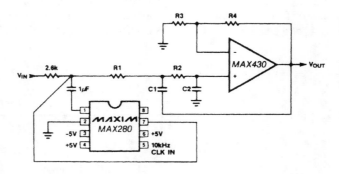

THE MAX430 IS CONNECTED AS A 2nd ORDER SALLEN AND KEY LOWPASS FILTER WITH A CUTOFF FREQUENCY EQUAL TO THE MAX280. THE ADDITIONAL FILTERING ELIMINATES ANY 10kHz CLOCK FEED THROUGH PLUS DECREASES THE WIDEBAND NOISE OF THE FILTER.

DC OUTPUT OFFSET (REFERRED TO A DC GAIN OF UNITY) = 5µV Max.

WIDEBAND NOISE (REFERRED TO A DC GAIN OF UNIT) = 60µ/V$_{RMS}$

OUTPUT FILTER COMPONENT VALUES					
DC GAIN	R3	R4	R1	R2	C1	C2
1	∞	0	14.3k	53.6k	0.1µF	0.033µF
10	3.57k	32.4k	4.6k	27.4k	0.1µF	0.2µF
101	0.324	32.4k	0.31k	16.9k	0.47µF	1µF

Figure 4–18.   MAX280 connected to provide lowpass cutoff at 100 Hz. (Maxim New Releases Data Book, 1992, p. 6-85)

Zero DC-Error Lowpass Filters

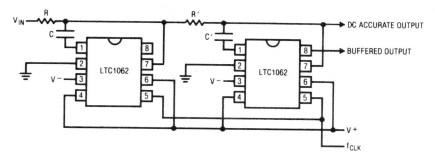

Figure 4-19. DC-accurate cascaded lowpass filter. (Linear Technology, Application Note 20, p. 7)

ratio of R'/R is about 117/1. 1/6.28RC should equal fc/1.57, and 1/6.28R'C' should equal fc/1.6. The clock should be 100 times the desired fc.

As an example, assume that the desired fc is 4.16 kHz. With pin 4 connected to +V, the divider ratio is at 100:1, so an external clock of 416 kHz is required. Using a response factor of fc/1.57, the result is 4160/1.57 = 2650. Then divide 1 by 2650 to find the desired product of 6.28RC, with a result of 1/2650 = 0.0003773. Assume a value of 1 k for R, then divide 0.0003773 by 6.28^{-3} for a C of 0.06 µF. The manufacturer recommends a standard R of 909 ohms and a C of 0.066 µF, for this particular circuit. Use a response factor of fc/1.6 to find R' and C'. The manufacturer recommends 107 k for R' and 574 pF for C'. You have the author's permission to verify these values for R' and C'.

4.13.4 Buffered Cascade Lowpass Filter (LTC1062)

Figure 4-20 shows two LTC1062 filters where the second input is taken from the buffered output of the first filter. This introduces a maximum DC error of 2 mV, over temperature, but now the values of R and R' can be similar in value, and the

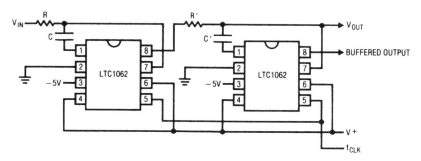

Figure 4-20. Buffered cascade lowpass filter. (Linear Technology, Application Note 20, p. 7)

80 SIMPLIFIED DESIGN OF FILTER CIRCUITS

passband-gain error is reduced to about 0.15 dB. The recommended ratio of R'/R is about 1.27/1. 1/6.26RC should equal fc/1.59, and 1/6.28R'C' should equal fc/1.64. The clock should be 100 times the desired fc.

As an example, assume that the desired fc is 4 kHz. With pin 4 connected to +V, the divider ratio is at 100:1, so an external clock of 400 kHz is required. The manufacturer recommends an R of 97.6 k, a C of 676 pF, an R' of 124 k, and a C' of 508 pF. Try these values on your calculator.

4.13.5 Low-Offset, 12th-Order Butterworth Lowpass Filter (LTC1062)

Figure 4–21 shows a 12th-order filter that uses two LTC1062s and a precision dual op amp. Figure 4–22 shows the frequency response for the following values: fc = 4 kHz, R = 59 k, C = 0.001 µF, R' = 5.7 k, C' = 0.01 µF, R1 = R2 = 39.8 k, C1 = 200 pF, C2 = 500 pF, f_{CLK} = 438kHz.

4.13.6 Notch Filter (LTC1062)

Figure 4–23 shows an LTC1062 and an LT1056 connected to form a notch filter, created from a lowpass filter. Figure 4–24 shows the frequency response for a 25-Hz notch filter using values based on the equations in Fig. 4–23. The optional R2/C2 at the LTC1062 output is used to minimize clock feedthrough. The 1/6.28R2C2 value should be 12 to 15 times that of the notch frequency.

4.13.7 Notch Filter (MAX280)

Figure 4–25 shows the MAX280 version of the circuit in Fig. 4–23. Figure 4–26 shows the phase-shift response for the circuit of Fig. 4–25, using various values of R and C. Use the equations to calculate the desired notch frequency. For example, to get a notch at 60 Hz, the clock frequency should be 7098 Hz. Divide this clock frequency by 163 to find the 1/6.28RC value. Then assume a standard value for C

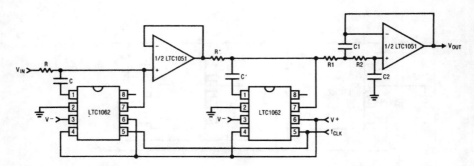

Figure 4–21. 12th-order Butterworth lowpass filter. (Linear Technology, Application Note 20, p. 8)

Zero DC-Error Lowpass Filters   81

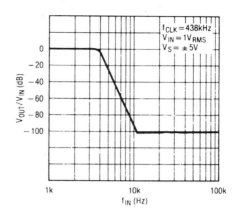

Figure 4–22.
Frequency response for 12th-order Butterworth lowpass filter. (Linear Technology, Application Note 20, p. 9)

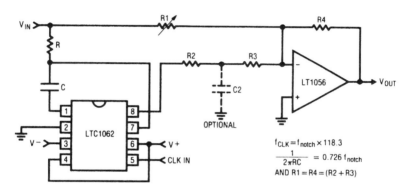

Figure 4–23.   Notch filter. (Linear Technology, Application Note 20, p. 9)

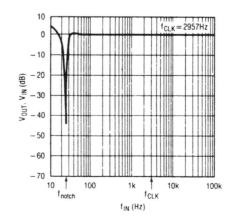

Figure 4–24.
Frequency response for notch filter. (Linear Technology, Application Note 20, p. 10)

Figure 4-25.
Notch filter using MAX280. (Maxim New Releases Data Book, 1992, p. 6-67)

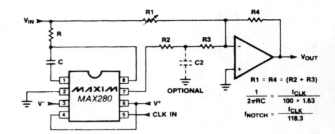

Figure 4-26.
Phase-shift response for notch filter. (Maxim New Releases Data Book, 1992, p. 6-87)

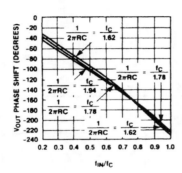

(0.1 μF, 0.01 μF, etc.) and calculate the value of R. Again, the 1/6.28R2C2 value should be 12 to 15 times that of the notch frequency.

4.13.8 Extended Notch Filter (LTC1062)

Figure 4–27 shows the LTC1062 combined with two sections of an LT1013 to form an extended notch filter where the output is a combination of lowpass and notch responses as shown in Fig. 4–28. The clock frequency is determined by the equation f_{CLK} = notch frequency × 47.3. Divide this clock frequency by 162 to find the 1/6.28RC value. Then assume a standard value for C (the manufacturer recommends 1 μF), and calculate the value of R.

For example, using our values, 60 × 47.3 = 2.84 kHz for the clock. 2.84 divided by 162 = 0.01753. 1 divided by 0.01753 = 57, and 57 divided by 6.28 = 9.09 k for R (assuming 1 μF for C). Note that the ratio of R6/R5 is 1.935, and R2 = R3 = R7, and R4 = R5 = 0.5R7.

4.13.9 Simple 5-Hz Filter (LTC1062)

Figure 4–29 shows the LTC1062 connected with solid-tantalum capacitors to form a 5-Hz lowpass filter. Notice that no external clock is required.

Zero DC-Error Lowpass Filters **83**

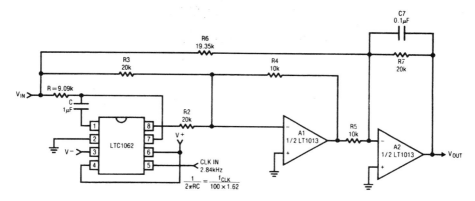

Figure 4-27. Extended notch filter. (Linear Technology, Application Note 20, p. 10)

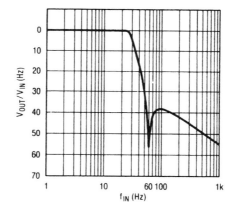

Figure 4-28.
Frequency response for extended notch filter. (Linear Technology, Application Note 20, p. 11)

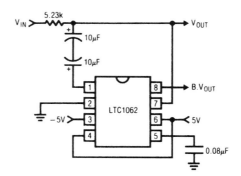

Figure 4-29.
Simple 5-Hz filter. (Linear Technology, Application Note 20, p. 11)

4.13.10 Clock-Sweepable Pseudo-Bandpass Notch Filter (LTC1062)

Figure 4–30 shows the LTC1062 connected as a simple clock-sweepable bandpass/notch filter. Figure 4–31 shows the frequency response for a clock frequency of 100 kHz and the various ratios of R1/R2. Figure 4–32 shows the variation of peak gain (Hop) and peak frequency (fp) versus different values of the R1/R2 ratio.

4.13.11 Selective Clock-Sweepable Bandpass Filter (LTC1062)

Figure 4–33 shows two LTC1062 ICs connected to form a clock-sweepable bandpass filter. Figure 4–34 shows the frequency response for the values given in Fig. 4–33.

4.13.12 Clock-Tunable Notch Filter (LTC1062)

Figure 4–35 shows an LTC1062 and an op amp (such as an LT1056) connected to form a notch filter. Figure 4–36 shows the frequency response for a 400-Hz notch filter that uses the following values and equations: R3 = R4 = R5 = 10 k, R1/R2 = a ratio of 1.234, $f_{CLK} \neq f_{notch} = 79.3/1$.

4.13.13 IC Filter with High Input Voltage (LTC1062)

Figure 4–37 shows how an LTC1062 can be connected to accommodate high input voltages outside the normal input common-mode range. The DC gain of the lowpass filter is R2/(R1 + R2). For maximum passband flatness, the paralleled combination of R1/R2 should be chosen as:

$$\frac{1}{6.28(R1 \parallel R2) \times C} = \frac{f_{cutoff}}{1.63}$$

where R1 ∥ R2 is equal or greater than 5 k. Notice that there is no need for an external op amp to buffer the divided-down input voltage. The internal buffer input (pin 7) performs this function.

4.13.14 IC Filter Operated from ±15-V Supplies (LTC1062)

Figure 4–38 shows an LTC1062 interfaced with an LT1013 op amp and operated from ±15-V supplies. The desired cutoff frequency is determined by the equation shown. A typical DC output is 300 mV.

4.13.15 IC Filter with Programmed Cutoff Frequencies (LTC1062)

Figure 4–39 shows an LTC used with a dual four-channel MUX (74HC4052) and an LT311 to provide four different cutoff frequencies (500, 250, 125, and 62.5 Hz).

Zero DC-Error Lowpass Filters 85

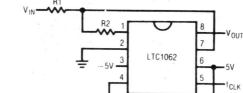

Figure 4–30.
Clock-sweepable pseudo-bandpass/notch filter. (Linear Technology, Application Note 24, p. 4)

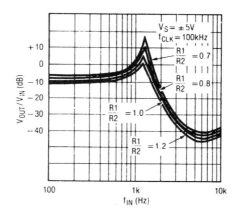

Figure 4–31.
Frequency response for clock-sweepable filter. (Linear Technology, Application Note 24, p. 5)

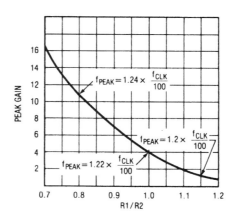

Figure 4–32.
Variations of peak gain and frequency for clock-sweepable filter. (Linear Technology, Application Note 24, p. 5)

86 SIMPLIFIED DESIGN OF FILTER CIRCUITS

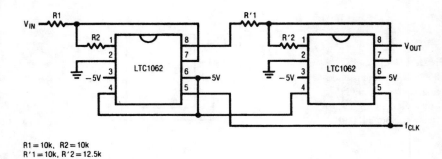

R1 = 10k, R2 = 10k
R'1 = 10k, R'2 = 12.5k

Figure 4–33. Selective clock-sweepable bandpass filter. (Linear Technology, Application Note 24, p. 5)

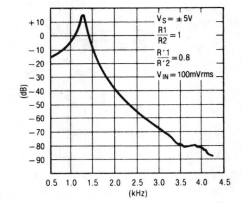

Figure 4–34. Frequency response for selective clock-sweepable filter. (Linear Technology, Application Note 24, p. 5)

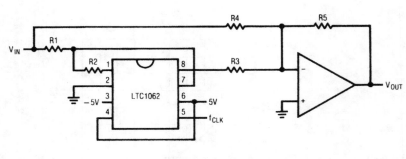

For simplicity use R3 = R4 = R5 = 10k;
$$\frac{R1}{R2} = 1.234, \quad \frac{f_{CLK}}{f_{notch}} = \frac{79.3}{1}$$

Figure 4–35. Clock-tunable notch filter. (Linear Technology, Application Note 24, p. 6)

Zero DC-Error Lowpass Filters 87

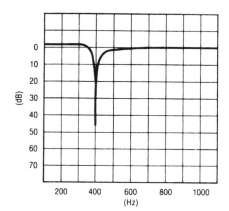

Figure 4-36.
Frequency response for clock-tunable notch filter. (Linear Technology, Application Note 24, p. 6)

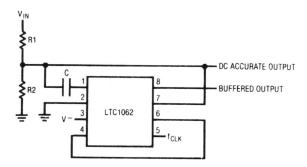

Figure 4-37.
IC filter with high input voltage. (Linear Technology, Application Note 24, p. 7)

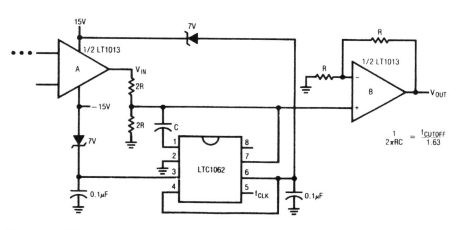

Figure 4-38. IC filter operated from ±15-V supplies. (Linear Technology, Application Note 24, p. 7)

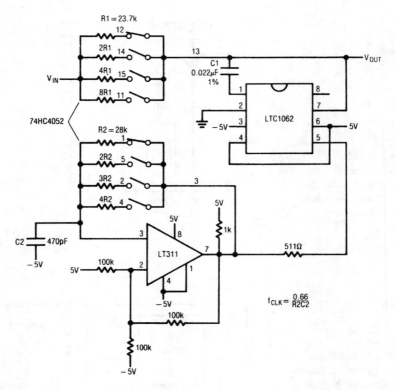

Figure 4–39. IC filter with programmed cutoff frequencies. (Linear Technology, Application Note 24, p. 8)

Notice that the clock frequency and the external R times C product are varied simultaneously so that $1/(6.28\,RC) = fc/1.64 = f_{clock}/164$.

4.13.16 First-Order DC-Accurate Lowpass Filter (MAX280)

Figure 4–40 shows a MAX280 filter and a CD4061 connected to provide a DC-accurate lowpass filter with selectable cutoff frequencies.

4.13.17 Lowpass Filter with Octave Tuning (MAX280)

Figure 4–41 shows three MAX280 filters connected to provide octave tuning with a single input clock. Use the equations of Fig. 4–27 to find the values of R and C. Figure 4–42 shows the amplitude response for the three filter outputs.

Zero DC-Error Lowpass Filters 89

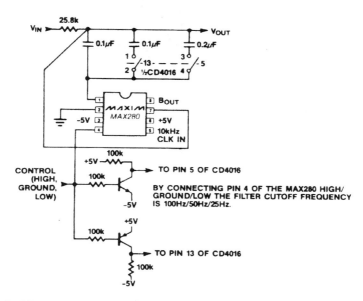

Figure 4-40. DC-accurate lowpass filter with selectable cutoff frequencies. (Maxim New Releases Data Book, 1992, p. 6-88)

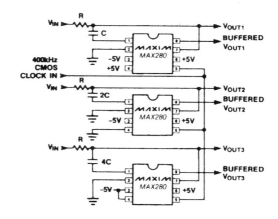

Figure 4-41.
Lowpass filter with octave tuning. (Maxim New Releases Data Book, 1992, p. 6-88)

Figure 4-42.
Amplitude response for lowpass filter with octave tuning. (Maxim New Releases Data Book, 1992, p. 6-88)

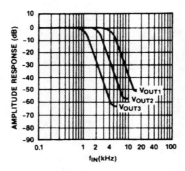

CHAPTER **5**

General-Purpose Lowpass Filters

This chapter is devoted to simplified design with general-purpose lowpass IC filters (the EXAR Corporation XR-1001/8). Each of these ICs is a 4th-order switched-capacitor lowpass filter providing 24-dB/octave (Butterworth) of rolloff outside of the passband. The series of 8 ICs provides for Butterworth, Bessel, or Chebyshev (0.5 or 0.1 dB of ripple) responses, depending on model number. Each filter response is available in 50:1 and 100:1 clock-to-corner ratios. All of the ICs have the same 8-pin pinout, so one can easily be substituted for the other, depending on application.

As in the case of other switched-capacitor filters, these ICs have the ability to tune the corner frequency of the filter response by selection of the external input clock. The ICs can also be used in a stand-alone mode, where an external resistor and capacitor set the input clock frequency. For additional precision, an external crystal can be used to set the corner frequency.

5.1 Filter Response and Applications

Figures 5–1 and 5–2 show the pinout and absolute-maximum ratings, respectively, for the eight ICs. Figure 5–3 shows the response/ripple and clock/corner characteristics for each IC. Note that the ICs are available in either plastic or ceramic. Figure 5–4 shows the electrical characteristics for all eight ICs. The minimum, typical, and maximum values shown in Fig. 5–4 were obtained using the test circuits of Figs. 5–5 and 5–6.

Note that Figure 3 referred to in the stand-alone operation portion of Fig. 5–4 is Figure 5–7 in this book, and that the stand-alone circuit is shown in Figure 5–8.

The XR-1001 is pin-for-pin compatible with the MF-4-100 and the XR1002 is pin-for-pin compatible with the MF-4-50. As shown in Fig. 5–3, these two ICs provide a Butterworth response where the passband must be maximally flat (such as instrumentation).

92 SIMPLIFIED DESIGN OF FILTER CIRCUITS

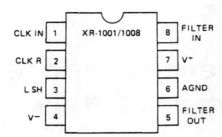

Figure 5–1.
Pinout for XR-1001/8.
(EXAR Corporation Databook, p. 5-127)

Figure 5–2.
Absolute Maximum Ratings for XR-1001/8.
(EXAR Corporation Databook, p. 5-127)

Power Supply (Single Supply)	14 V
Input Signal Level	V+ −0.7 to V− + 0.7 V
Power Dissipation (Package Limitation)	
Ceramic Package	385 mW
Derate Above T_A = 25°C	5 mW/°C
Plastic Package	300 mW
Derate Above T_A = 25°C	6 mW/°C
Storage Temperature Range	−65°C to +150°C

Part Number	Package	Response/Ripple	f_{clock}/f_{corner}	Operating Temperature
XR-1001CP/CN	Plastic/Ceramic	Butterworth	100:1	0°C to 70°C
XR-1002CP/CN	Plastic/Ceramic	Butterworth	50:1	0°C to 70°C
XR-1003CP/CN	Plastic/Ceramic	Bessel	100:1	0°C to 70°C
XR-1004CP/CN	Plastic/Ceramic	Bessel	50:1	0°C to 70°C
XR-1005CP/CN	Plastic/Ceramic	Chebyshev (0.1dB)	100:1	0°C to 70°C
XR-1006CP/CN	Plastic/Ceramic	Chebyshev (0.1 dB)	50:1	0°C to 70°C
XR-1007CP/CN	Plastic/Ceramic	Chebyshev (0.5 dB)	100:1	0°C to 70°C
XR-1008CP/CN	Plastic/Ceramic	Chebyshev (0.5 dB)	50:1	0°C to 70°C

Figure 5–3. Response/Ripple and Clock/Corner Characteristics for XR-1001/8. (EXAR Corporation Databook, p. 5-127)

The XR-1003 (Fig. 5–9) and XR1004 (Fig. 5–10) have a Bessel response, which is ideal for telecommunications and modem applications (where phase distortion might affect performance).

The XR-1005 through XR-1008 (Figs. 5–11 through 5–14) have a Chebyshev response, where rolloff outside of the passband is steeper and attenuates out-of-band signals greater than the Bessel or Butterworth response. However, as discussed, the ripple inside the Chebyshev band is greater than for Bessel or Butterworth.

ELECTRICAL CHARACTERISTICS

Test Conditions: $V^+ = 5$ VDC, $V^- = -5$ VDC, $f_{CLOCK} = 1$ MHz, $R_{Load} = 1$ Megohm, $C_{Load} = 40$ pF, $T_A = 25°C$, unless specified otherwise.

SYMBOL	PARAMETER	MIN	TYP	MAX	UNIT	CONDITION
GENERAL CHARACTERISTICS						
V_{DD}	Supply Voltage Single Supply Split Supply	4.5 +2.25		11.0 5.5		VDC Referenced to V_{SS} VDC (Pin 4)
V_{SS}	Supply Voltage Split Supply	-5.5		-2.25		VDC Reference to AGND (Pin 6)
I_{DD}	Supply Current Single Supply Split Supply Split Supply		2.5 2.5 1.50	3.5 3.5 2.25	mA mA mA	$V_{DD} = 10.0$ VDC $V_{DD} = -5$ VDC $V_{DD} = 2.25$ VDC
I_{SS}	Supply Current Split Supply Split Supply	-3.5 -2.25	-2.5 -1.5		mA mA	$V_{DD} = 5.0$ VDC, $V_{SS} = -5.0$ VDC $V_{DD} = +2.25$ VDC, $V_{SS} = -2.25$
FILTER CHARACTERISTICS						
$f_{CLOCKMAX}$	Upper Clocking Freq. Limit	1.0	1.5		MHz	
$f_{CLOCKMIN}$	Lowest Practical Clock		100 50		Hz Hz	For 1001,1003,1005,1007 For 1002,1004,1006,1008
	Gain at Corner Frequency	-3.5	-3	-2.5	dB	
$A_{2f_{corner}}$	Attenuation at 2 Times The Corner Frequency	23 11 30 33	24.6 13 31 34	26 14 33 36	dB dB dB dB	For 1001,1002 For 1003,1004 For 1005,1006 For 1007,1008
V_{out}	Maximum Output Signal	8			Vpp	Input = ±4.2 VDC
$e_{n\ out}$	Output Noise		0.5		mvrms	From 1 Hz to 25 kHz
S/N	Signal-to-Noise Ratio		84		dB	
THD	Total Harmonic Distortion		0.1		%	$V_{in} = 2.4$ Vrms, $f_{in} = 1$ kHz
V_{OS}	Output Offset Voltage (DC)	-0.4		+0.4	VDC	$f_{clock} = 1$ MHz
	Clock Feedthrough		50		mvpp	
I_{OSS}	Output Short Circuit Current Source Sink	-60	-50 30	50	mA mA	See Note #1
	Temperature Coefficient Of f_{corner}		±35		ppm/°C	From -40°C to +85°C ; not tested in production
	Passband Gain For 1001,1002	-0.3	0	+0.3	dB	Tested at 3 and 5 kHz for 1001,1003,1005,1007
	For 1003,1004	-1.0	-0.1	+0.2	dB	Tested at 6 and 10 kHz for 1002,1004,1006,1008
	For 1005,1006	-0.4	-0.1	+0.3	dB	
	For 1007,1008	-0.7	-0.2	+0.3	dB	

#1 Caution should be used so that the power dissipation does not exceed the package limitation.

Figure 5–4. Electrical Characteristics for XR-1001/8. (EXAR Corporation Databook, pp. 5-218, 5-129) (*Figure continued on next page.*)

94 SIMPLIFIED DESIGN OF FILTER CIRCUITS

	Passband Gain				dB	
	For 1001,1002	-0.7	-0.06	-0.0	dB	Tested at 7.5 kHz for 1001,1003,1005,1007
	For 1003,1004	-2.0	-1.5	-1.0	dB	Tested at 15 kHz for 1002,1004,1006,1008
	For 1005,1006	-0.3	-0.1	+0.3	dB	
	For 1007,1008	0.7	-0.2	+0.3	dB	
FILTER CHARACTERISTICS: V^+ = 2.25, V^- = -2.25 VDC						
$f_{CLOCKMAX}$	Upper Clocking Freq. Limit	0.25	0.5		MHz	
$f_{CLOCKMIN}$	Lower Practical Clock		100		Hz	For 1001,1003,1005,1007
			50		Hz	For 1002,1004,1006,1008
$A_{2f_{corner}}$	Attenuation at 2 Times	23	24.6	26	dB	For 1001,1002
	The Corner Frequency	11	13	14	dB	For 1003,1004
		30	31	33	dB	For 1005,1006
		33	34	36	dB	For 1007,1008
	Maximum Output Signal	3	4		Vpp	V_{in} = ±2.0 VDC
S/N	Signal-to-Noise Ratio		76		dB	
V_{OS}	DC Offset Voltage	-0.4	±0.05	+0.4	VDC	
LOGIC INPUT & LOGIC OUTPUT TESTS: V = ±2.25 VDC and V = ±5 VDC, Pin 3 tied to V_{SS}						
	Schmitt Trigger Input					
V_T+	Positive Going Threshold	0.6	1.3	2.0	V	V = ±5.0 VDC
	Voltage	0.0	0.55	+1.1	V	V = ±2.25 VDC
V_T-	Negative Going Threshold	-1.4	-0.7	0.0	V	V = ±5.0 VDC
	Voltage	-0.6	-0.2	+0.4	V	V = ±2.25 VDC
$V_T+ - V_T-$	Hysteresis	0.8	2.1	2.9	V	V = ±5.0 VDC
		0.2	0.75	1.3	V	V = ±2.25 VDC
V_{OH}	Output High Voltage	4.5			V	V = ±5.0 VDC
		2.03			V	V = ±2.25 VDC, I_o = 400µA
V_{OL}	Output Low Voltage	1			V	V = ±5.0 VDC
		0.5			V	V = ±2.25 VDC, I_o = 400µA
I_{OS}	Output Sink Current	2.5	5.0		mA	V = ±5.0 VDC
		0.65	1.3		mA	V = ±2.25 VDC
	Output Source Current	3.0	6.0		mA	V = ±5.0 VDC
		0.75	1.5		mA	V = ±2.25 VDC
TTL CLOCK INPUT: V = ±5.0 VDC, Pin 3 tied to 0 VDC						
V_{IL}	Input Low Voltage		0.8		V	
V_{IH}	Input High Voltage		2.8		V	
STAND ALONE OPERATION: R = 5.0 kilohms and C = 120 pF for f_o = 985 kHz						
f_o	Frequency Accuracy of Oscillator	-15	±4	+15	%	Measured at Pin 2. Error increases at lower f_o (<500kHz). See Figure 3.

Figure 5-4. (Continued.)

General-Purpose Lowpass Filters **95**

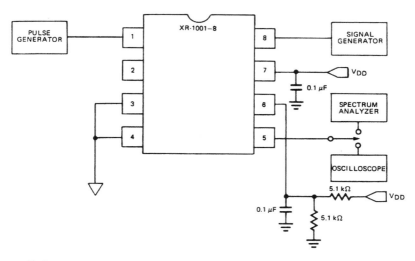

Figure 5-5. Single-supply test connections for XR-1001/8. (EXAR Corporation Databook, p. 5-130)

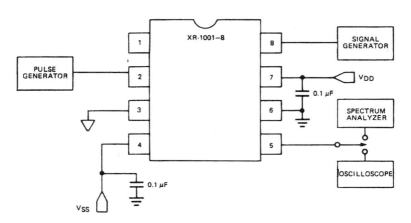

Figure 5-6. Split-supply test connections for XR-1001/8. (EXAR Corporation Databook, p. 5-130)

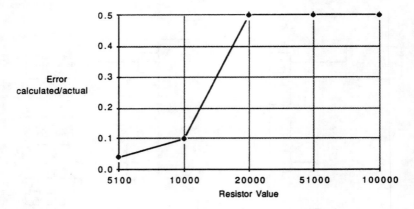

Figure 5-7. Error versus timing resistor for XR-1001/8. (EXAR Corporation Databook, p. 5-132)

Figure 5-8. Stand-alone (RC) connections for XR-1001/8. (EXAR Corporation Databook, p. 5-132)

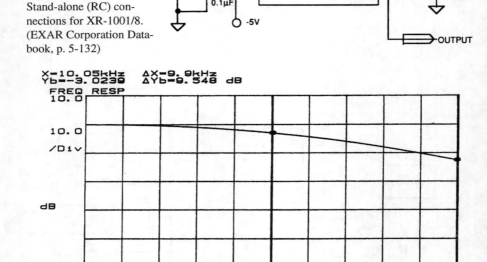

Figure 5-9. Bessel response for XR-1003. (EXAR Corporation Databook, p. 5-133)

General-Purpose Lowpass Filters   **97**

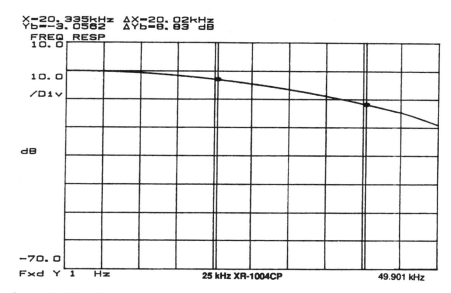

Figure 5-10. Bessel response for XR-1004. (EXAR Corporation Databook, p. 5-133)

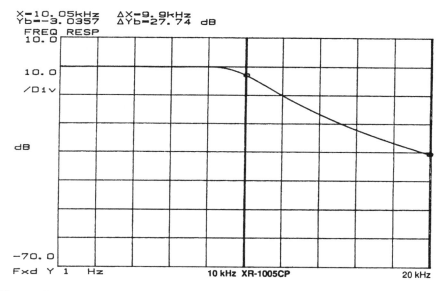

Figure 5-11. Chebyshev response for XR-1005. (EXAR Corporation Databook, p. 5-134)

98 SIMPLIFIED DESIGN OF FILTER CIRCUITS

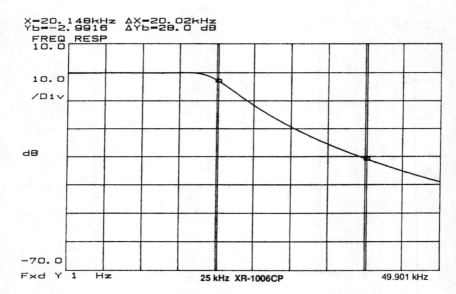

Figure 5-12. Chebyshev response for XR-1006. (EXAR Corporation Databook, p. 5-134)

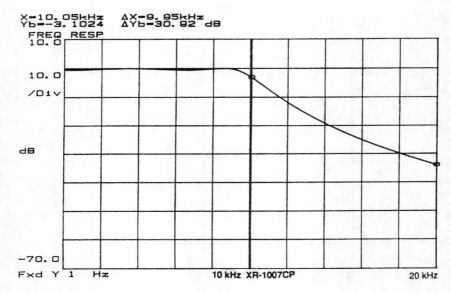

Figure 5-13. Chebyshev response for XR-1007. (EXAR Corporation Databook, p. 5-135)

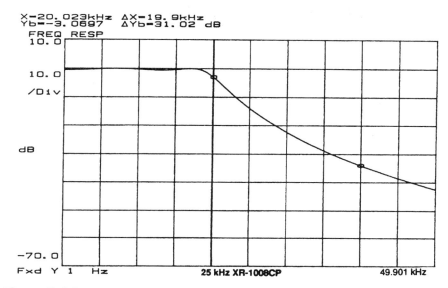

Figure 5-14. Chebyshev response for XR-1008. (EXAR Corporation Databook, p. 5-135)

5.2 Pin Descriptions for XR-1001/8

The following is a summary of the pin functions, together with notes for simplified design using the XR-1001/8 ICs.

Pin 1 (CLK IN) is the clock input. The clock should be MOS-level such as used for sampling and switching. For best results, the clock should have a 50-percent duty cycle. As an alternate clock source, use a crystal connected from pin 1 to pin 2. This results in a self-contained oscillator operating at the exact crystal frequency. Connect a 10-M resistor in parallel with the crystal. If the exact frequency is not critical, use the resistor and capacitor for stand-alone operation shown in Fig. 5–8. As shown, the capacitor is tied from pin 1 to ground, and the resistor is connected between pin 1 and pin 2. With any configuration, the maximum recommended clock frequency is 1 MHz.

Pin 2 (CLK R) is for the clock resistor. The signal at pin 2 is the inverted output of the signal at pin 1. The main purpose for pin 2 is when the IC is used as a crystal-controlled filter, or for stand-alone RC operation as shown in Fig. 5–8. The frequency of oscillation using the RC circuit is equal to 1/(1.7 times RC). Pin 2 can also be used as a TTL-level input for external clock signals. (If the external clock is MOS-level, use pin 1.)

Pin 3 (L SH) is for level shift. The input is used to set the logic-zero point of the clock input. For MOS-level clocks, tie pin 3 to VSS. For TTL-level clock inputs, tie pin 3 to ground (for a split supply). For single-supply operation, a MOS-level clock is recommended. When a crystal and parallel resistor, or a resistor and capacitor, are used to create an oscillator (no external clock), tie pin 3 of VSS. See Figs. 5–5, 5–6, and 5–8 for typical connections.

Pin 4 (V–) is the negative supply terminal (VSS). Pin 4 should have substantial decoupling to prevent noise at the filter output. A minimum of 0.1-µF capacitance to ground is strongly recommended. Make the decoupling capacitor connection as close to the IC as possible.

The range of operation for dual supplies is from –2.5 V to –5 V. The ICs can also be operated from ground positive. In this case, the VSS pin 4 is tied to the analog ground of the circuit. The quality of this ground in very important. Series inductance must be at an absolute minimum. (Use the shortest possible lead.)

Pin 5 (FILTER OUT) is the lowpass filter output. The recommended load is 10 k, or larger, at pin 5.

Pin 6 (AGND) is the analog ground. Pin 6 should be tied to the analog ground of the external circuit. The filter output (pin 5) of the IC will swing around the potential at pin 6. For single-supply operation, pin 6 must be externally biased at a level one-half of the VDD voltage. Use an external bias circuit for pin 6 similar to that shown in Fig. 5–5. A 0.1-µF capacitor is the minimum value for capacitance at pin 6 when single-supply operation is used. Pin 6 should be grounded as shown in Fig. 5–6 for split-supply operation.

Pin 7 (V+) is VDD. The positive supply (either single or split) is connected to pin 7. The manufacturer recommends a 1-µF capacitor in parallel with the 0.1-µF capacitor shown at pin 7 in both Figs. 5–5 and 5–6. The additional capacitance is needed to decouple the positive supply.

The range of operation for the IC is from +5 to +10 V when single-supply is used. For split-supply (also known as dual-supply), the range is from +2.5 V to +5 V.

Pin 8 (FILTER IN) is the lowpass filter input. The input signal should be biased to mid-supply before application to pin 8. As usual, to prevent aliasing, the input frequency must be less than one-half the clock frequency.

CHAPTER **6**

General-Purpose Elliptic Lowpass Filters

This chapter is devoted to simplified design with elliptic lowpass IC filters (the EXAR Corporation XR-1015/16). These ICs are switched capacitor, so the passband is set by the clock frequency. The XR-1015 and XR-1016 are 7th-order devices that also provide synchronized sampled inputs and outputs. This feature permits the ICs to be cascaded without the need for additional sample/hold circuits. (Refer to Chapter 7 for additional information on cascading.)

The XR-1015 is an 8-pin device that can operate from +3, –2 V to ±5 V, and can also be biased to operate with a single +5 V to +10 V supply. The XR-1015 is pin-for-pin compatible with the Reticon R5609, but with the added advantage of operating from a single supply. The clock-to-corner ratio of the XR-1015 is fixed at 100:1.

The XR-1016 is a 14-pin device that provides two uncommitted op amps for use as a reconstruction filter, anti-aliasing filter, or for additional filter gain. As in the case of the XR-1015, the XR-1016 provides a clock output (with rail-to-rail output). The XR-1016 clock-to-corner ratio can be either 100:1 or 50:1 as needed. The output clock can be used to strobe an analog-to-digital (A/D or ADC) converter or to synchronize any additional circuits in the system.

6.1 Filter Characteristics

Figures 6–1 and 6–2 show the pinout and absolute-maximum ratings, respectively, for the ICs. Figure 6–3 shows the electrical characteristics. The minimum, typical, and maximum values shown in Fig. 6–3 were obtained using the test circuits of Figs. 6–4 through 6–7. Note that Figures 1 through 4 referred to in the characteristics of Fig. 6–3 are Figures 6–4 through 6–7 in this book.

Figures 6–8 and 6–9 show the amplitude response and group-delay response, respectively. Figures 6–10 and 6–11 show the typical passband ripple and output noise, respectively. As shown in Fig. 6–8, any signal from the –3-dB pint of the low-

102 SIMPLIFIED DESIGN OF FILTER CIRCUITS

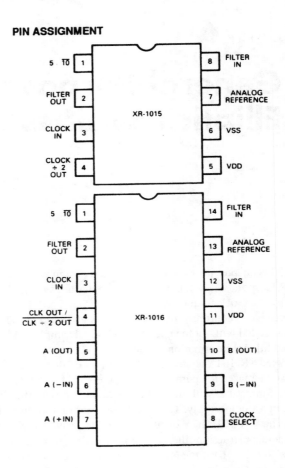

Figure 6–1.
Pinouts for XR-1015/16.
(EXAR Corporation Databook, 1992, p. 5-37)

Figure 6–2.
Absolute Maximum Ratings for XR-1015/16.
(EXAR Corporation Databook, 1992, p. 5-37)

Power Supply	14V
Input Signal Level	V+ +0.3 to V− ±0.3V
Power Dissipation — XR-1016 (Package Limitation)	
Ceramic Package	1000mW
Derate Above T_A = +25°C	6mW/°C
Plastic Package	800mW
Derate Above T_A = +25°C	7mW/°C
Power Dissipation — XR-1015	
Ceramic Package	385mW
Derate Above T_A = +25°C	8.3mW/°C
Plastic Package	300mW
Derate Above T_A = +25°C	8.3mW/°C
Storage Temperature	−55°C to +150°C

General-Purpose Elliptic Lowpass Filters

Test Conditions: V+ = 5 VDC, V− = −5 VDC, f_{CLOCK} = 2MHz, R_L = 1 MΩ, C_L = 40pF, T_A = 25°C, unless otherwise noted.

SYMBOL	PARAMETERS	XR-1015 MIN	XR-1015 TYP	XR-1015 MAX	XR-1016 MIN	XR-1016 TYP	XR-1016 MAX	UNITS	CONDITIONS
GENERAL CHARACTERISTICS									
	Supply Voltage								
	Single Supply	5		10.5				V	See Figure 1
					5		10.5	V	See Figure 3
	Split Supply	+3, −2		15.25				V	See Figure 2
					+3, −2		±5.25	V	See Figure 4
	Supply Current								
	Single Supply		10	11				mA	See Figure 1
						10	11	mA	See Figure 3
	Split Supply								
	Positive		10	12				mA	See Figure 2
	Negative		10	12				mA	
	Positive					10	12	mA	See Figure 4
	Negative					10	12	mA	
FILTER SECTION									
f_{CLOCK}	Upper Frequency Limit	2	2.5					MHz	See Figure 2
					2	2.5		MHz	See Figure 4
$f_{CLOCKMIN}$	Lowest Practical		1					kHz	See Figure 2
						1		kHz	See Figure 4
	Input Impedance								
	Pin 8		1					MΩ	See Figure 2, f_{CLOCK} = 1MHz
	Pin 14					1		MΩ	See Figure 4. f_{CLOCK} = 1MHz
tpw	Minimum f_{CLOCK} Pulse Width	200						ns	See Figure 2
					200			ns	See Figure 4
THD	Total Harmonic Distortion								Vin = 2 Vpp
			0.02%			0.02%			f_{CLOCK} = 500kHz
			0.1%			0.1%			f_{CLOCK} = 2MHz
	Clock Feed through		30			30		mVpp	
V_{INMAX}	Maximum Input Voltage		8			8		Vpp	Above which distortion increases.
	Corner Freq. Accuracy		±0.5%	±1%		±0.5%	±1%		f_{CLOCK} = 2MHz
A_V	Passband Gain	−0.5	0	+0.5	−0.5	0	+0.5	dB	Tested at f_{in} = 293Hz, 3.9kHz, 8.6kHz, 12.1kHz
		−1	0	+1	−1	0	+1	dB	Tested at f_{in} = 15.3kHz, 17.1kHz
	Ripple Passband		±0.1			±0.1		dB	f_{CLOCK} = 500kHz
			±0.5	1		±0.5	1	dB	f_{CLOCK} = 2MHz
V_{OS}	Voltage Offset	−0.5	−0.2	+0.5	−0.5	−0.2	+0.5	VDC	
	Output Noise		0.6			0.6		mVrms	1Hz–20Hz, See Figure 8
OPERATIONAL AMPLIFIER									
	Unity Gain Bandwidth		1.2			1.2		MHz	
CMRR	Common Mode Rejection Ratio (2 Vpp Input)		50			50		dB	
V_{IO}	Input Offset Voltage	−30		30	−30		30	mV	

Figure 6–3. Electrical Characteristics for XR-1015/16. (EXAR Corporation Databook, 1992, pp. 5-36, 5-37)

104 SIMPLIFIED DESIGN OF FILTER CIRCUITS

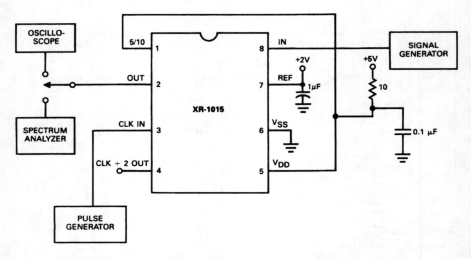

Figure 6–4. Test circuit. (5-V operation) for XR-1015. (EXAR Corporation Databook, 1992, p. 5-40)

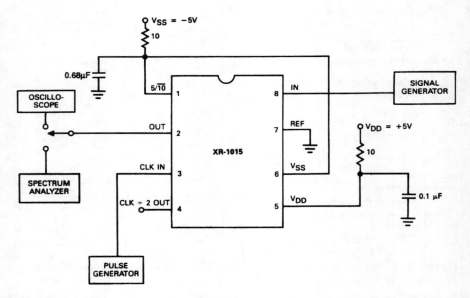

Figure 6–5. Test circuit. (10-V operation) for XR-1015 (EXAR Corporation Databook, 1992, p. 5-40)

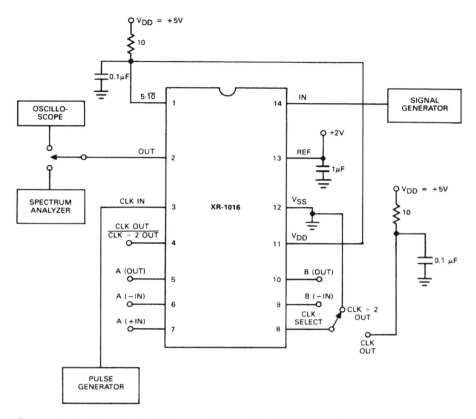

Figure 6-6. Test circuit. (5-V operation) for XR-1016 (EXAR Corporation Databook, 1992, p. 5-41)

pass response to one-half of the sampling frequency (one-quarter of the clock frequency) will be attenuated by about 75 dB, referenced to the passband. This allows use with A/D converters to prevent the A/D from aliasing signals that are above one-half of the sampling frequency of the A/D converter. A simple second-order active filter can be used in front of the XR-1015 or XR-1016 if input signals will be above the Nyquist frequency of the filters. When the ICs are used with digital-to-analog converters an active filter can be used at the output of the XR-1015 or XR-1016 to prevent the sampling frequency from causing problems with other stages of the system.

106 SIMPLIFIED DESIGN OF FILTER CIRCUITS

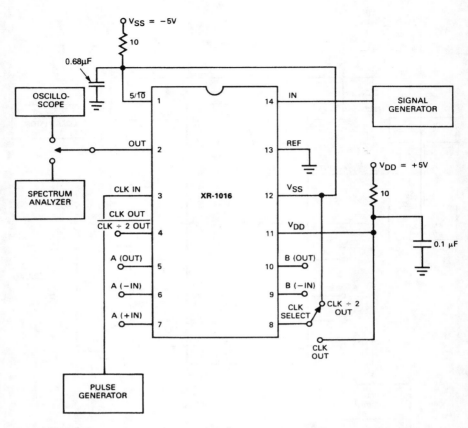

Figure 6–7. Test circuit. (10-V operation) for XR-1016 (EXAR Corporation Databook, 1992, p. 5-42)

6.2 Applications Requirements

The XR-1015/16 are fabricated in P-well CMOS. This uses an N-substrate and requires that VDD be applied first before VSS to prevent latchup of the IC. Also, the input signals, including the input clock, should not be at any voltage above the power-supply level (to prevent latchup). The input signal should not have any traces or wires near the clock or other system clocks (always a good idea!). The same is true of the output. This will help to reduce the clock feedthrough and provide measurements equal to the datasheet values.

The elliptic response of the XR-1015/16 is called an *equal-ripple response* where the ripple in the stopband is approximately the same as the ripple in the passband (Figs. 6–8 and 6–10). The methods used to get equal ripple (and the steep rolloff) do cause some change in the linearity of the group-delay of these elliptic filters.

General-Purpose Elliptic Lowpass Filters **107**

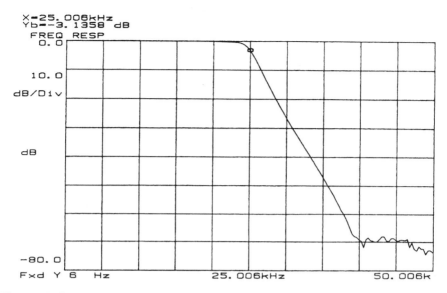

Figure 6–8. Amplitude response for XR-1015/16. (EXAR Corporation Databook, 1992, p. 5-43)

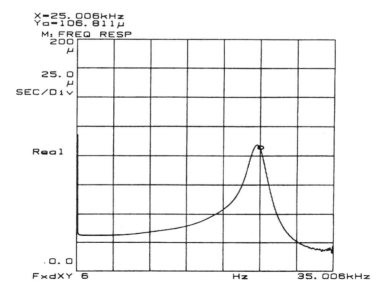

Figure 6–9. Group-delay response for XR-1015/16. (EXAR Corporation Databook, 1992, p. 5-43)

108 SIMPLIFIED DESIGN OF FILTER CIRCUITS

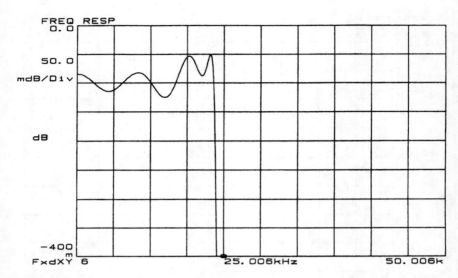

Figure 6–10. Typical passband ripple for XR-1015/16. (EXAR Corporation Databook, 1992, p. 5-44)

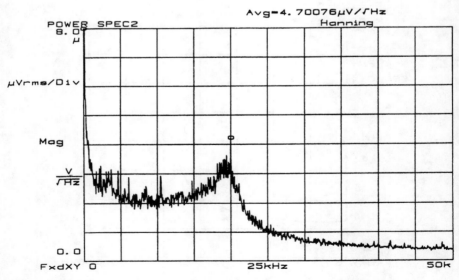

Figure 6–11. Typical output noise for XR-1015/16. (EXAR Corporation Databook, 1992, p. 5-44)

General-Purpose Elliptic Lowpass Filters 109

Figure 6-12. Possible input-logic thresholds for various combination of VDD/VSS. (EXAR Corporation Databook, 1992, p. 5-45)

V_{DD}/V_{SS}	Level at Pin 1	Logic Decision Level
+5/−5 VDC	−5 VDC	1.8 VDC
	+5	−1.8 VDC
+5/0 VDC	0	3.7 VDC
	+5	1.8 VDC
+2.5/−2.5 VDC	−2.5	1.2 VDC
	+2.5	−0.7 VDC
+10/0 VDC	0	6.8 VDC
	+10	3.2 VDC

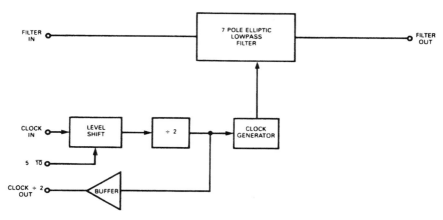

Figure 6-13. Internal functions of XR-1015. (EXAR Corporation Databook, 1992, p. 5-47)

The rapid change in group delay occurs near the corner frequency as shown in Fig. 6–9.

Because the XR-1015/16 are sample-data filters, in that they divide the continuous time signal into an amount of charge at a given time, certain limitations are placed on the signals to be filtered. The frequency of the signal applied at the filter input must have a period long enough so that at least two samples are taken during the period. This is true even if the signal to be filtered is in the stopband response of the filter.

The reason for this restriction is that it would take a minimum of two samples of the input signal frequency to establish the period of the signal, as well as an approximation of the amplitude. If this sampling criteria is not followed, the filter output will be an aliased signal of the input (because the period, and thus the frequency, would not be known).

110   SIMPLIFIED DESIGN OF FILTER CIRCUITS

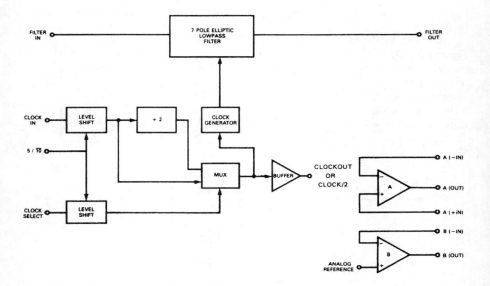

Figure 6-14.   Internal functions of XR-1016. (EXAR Corporation Databook, 1992, p. 5-47)

This aliasing problem can be eliminated with a simple second-order filter at the input. With the XR-1016, op amp A could be used to create the input filter. The precise location of the input-filter corner frequency is not critical, so precision resistors and capacitors are not needed.

6.3 Pin Descriptions for XR-1015/16

The following is a summary of the pin functions, together with notes for simplified design using the XR-1015/16. Because the XR-1016 has more functions (and pins) than the XR-1015, the functions are not necessarily the same for a given pin.

Pin 1 (5/10) for XR-1015/16. This pin controls the reference level of the internal level-shifter circuits of the ICs. In turn, this sets the point at which the IC considers the digital control input (not the input signal to be filtered) a logic 1 or a logic 0. When pin 1 is tied at VSS, the decision level is two-thirds of the sum of the VDD and VSS levels relative to VSS. When pin 1 is tied high (VDD) the decision level is set for two-thirds of the sum of VDD and VSS relative to VDD. Figure 6–12 shows some of the possibilities for input-logic thresholds. Figures 6–13 and 6–14 show the internal functions of the XR-1015 and XR-1016, respectively.

General-Purpose Elliptic Lowpass Filters **111**

Pin 2 (FILTER OUT) for XR-1015/16. This is the filter output and is designed to drive a 10-k load. The output signal will be centered around the voltage set by the ANALOG REFERENCE pin.

Pin 3 (CLOCK IN) for XR-1015. The input clock is applied at pin 3 and controls the position of the filter corner frequency using the ratio fclock/fcenter = 100:1. The logic-threshold level required at pin 3 is set by the status of pin 1 as shown in Fig. 6–12.

Pin 3 (CLOCK IN) for XR-1016. The input clock is applied at pin 3, and controls the position of the filter corner frequency. The XR-1016 has an internal divider that provides either a clock-to-corner ratio of either 100:1 or 50:1. This is controlled by pin 8 (CLOCK SELECT). If pin 8 is low, fclock/fcorner = 100:1.

Pin 4 (CLOCK/2) for XR-1015/16. This output is at the same frequency as the sampling frequency and can be used to synchronize an A/D converter to the filter output. The falling edge of the CLOCK/2 output is the edge that the output (pin 2) should be sampled to ensure that the output has settled.

Pin 5 (VDD) for XR-1015. This is the positive supply terminal. See Fig. 6–12 for typical VDD/VSS values. Refer to Pin 11 for XR-1016.

Pin 6 (VSS) for XR-1015. This is the negative supply terminal. See Fig. 6–12 for typical VDD/VSS values. Refer to Pin 12 for XR-1016.

Pin 5 (A OUT) for XR-1016. This is the output of op amp A, and is provided for creation of additional filtering (or some similar function). The output can drive a typical 10-k load.

Pin 6 (A-IN) for XR-1016. This is the negative or inverting input of op amp A, and is a CMOS-gate input with virtually infinite input impedance.

Pin 7 (A +IN) for XR-1016. This is the positive or non-inverting input of op amp A, and is a CMOS-gate input with virtually infinite input impedance.

Pin 7 (ANALOG REFERENCE) for XR-1015. Pin 7 controls the level at which the analog signals will be referenced. If equal split supplies are used, pin 7 should be tied to ground, as shown in Fig. 6–5. If unequal supplies (or a single supply) are used, then pin 7 should be connected to a voltage that is one-half of the algebraic sum of the two supplies. This can be done using a high-resistance voltage divider connected between VDD and VSS. Because pin 7 is used as the analog ground within the XR-1015, use a decoupling capacitor (of at least 0.47 μF) from pin 7 to ground as shown in Fig. 6–4. (Increase capacitance as clock frequency decreases, and vice versa.)

Pin 8 (FILTER IN) for XR-1015. This is the lowpass filter input. The input impedance is about 4 M at a 1-MHz clock frequency. As always, to prevent aliasing, the input frequency must be less than one-half the clock frequency.

Pin 8 (CLOCK SELECT) for XR-1016. This pin selects the clock-to-corner ratio of the filter response. When pin 8 is at logic 0 (Fig. 6–6), the filter has a clock-to-corner ratio of 100:1. A 50:1 clock-to-corner ratio is selected when pin 8 is at a logic 1.

Pin 9 (B -IN) for XR-1016. This is the negative or inverting input of op amp B. Note that the positive or non-inverting input of op amp B is tied internally to the ANALOG REFERENCE pin (Fig. 6–14).

Pin 10 (B OUT) for XR-1016. This is the output of op amp B.

Pin 11 (VDD) for XR-1016. This is the positive supply terminal, and is the same as pin 5 for the XR-1015. The range of positive supply voltages is from +2.5 V to +5 V when using dual supplies of equal voltage. If a single supply is used, the range is from +5 V to +10 V. It is recommended that a decoupling capacitor be connected from pin 11 to ground as shown in Figs. 6–4 through 6–7. This will minimize output noise. Figure 6–11 shows typical output noise response up to 50 kHz. (Note that the output noise value given in Fig. 6–3, and Figure 8, refers to Fig. 6–11 in this book.) The decoupling capacitor should be located as close to the VDD pin as is practical. Also, it is recommended that a 10-ohm resistor be connected between pin 11 and the positive supply, as shown in Figs. 6–4 through 6–7. The resistor will help prevent supply noise from entering the filtered output.

Pin 12 (VSS) for XR-1016. This is the negative supply terminal, and is the same as pin 6 for the XR-1015. The range of negative supply voltages is from –5 V to 0 V. Again, a decoupling capacitor and a 10-ohm series resistor are recommended for VSS as shown in Figs. 6–5 and 6–7 (when split supplies are used). For single-supply operation (Figs. 6–4 and 6–6) connect VSS directly to ground.

Pin 13 (ANALOG REFERENCE) for XR-1016. This is the same as pin 7 for the XR-1015. In the case of a single supply (or unequal supplies), connect pin 13 to a voltage that is one-half of the algebraic sum of the two supplies (Figs. 6–4 and 6–6). If equal split supplies are used, connect pin 13 directly to ground.

Pin 14 (FILTER IN) for XR-1016. This is the lowpass filter input, and is the same as for pin 8 of the XR-1015.

CHAPTER **7**

Tabular Design of Bandpass Filters

This chapter is devoted to simplified (tabular) design of bandpass filters. Two approaches (both developed by Linear Technology) are described. Each method uses switched-capacitor filters connected in cascade. One method uses cascaded non-identical 2nd-order ICs to form filters with Butterworth and Chebyshev responses. The second method uses identical 2nd-order bandpass sections in cascade.

7.1 Using the Tables

The design procedures in this chapter are based on the use of the Linear Technology tables shown in Figs. 7–1 through 7–4. These tables were derived from textbook filter theory to be used with Linear Technology IC filters (the LTC1059, 1060, 1061, and 1064). The tables can be applied to this family of IC filters provided that the design Q be kept low (less than 20), and the tuning resistors are at least 1 percent tolerance.

When the Q is kept less than 20, it is possible to design bandpass filters (using the Linear Technology switched-capacitor ICs) with almost textbook performance. When the Q must be higher, tuning should be avoided and the "A" versions of the LTC1059, 60, 61, or 64 should be specified. Also, resistor tolerances of better than 1 percent are necessary for the higher-Q filters.

7.1.1 Butterworth and Chebyshev Tables

The table of Fig. 7–1 is used to design Butterworth bandpass filters with a generalized response as shown in Fig. 7–5. The tables of Figs. 7–2, 7–3, and 7–4 are used for design of bandpass filters with a Chebyshev response similar to that of Fig. 7–6. As shown in Figs. 7–5 and 7–6, the tables provide responses that are geometrically

113

114 SIMPLIFIED DESIGN OF FILTER CIRCUITS

f_{oBP} (Hz)	f_{oBP}/BW (Hz)	f_{o1} (Hz)	f_{o2} (Hz)	f_{o3} (Hz)	f_{o4} (Hz)	f_{-3dB} (Hz)	f_{-3dB} (Hz)	Q1 = Q2	Q3	K	f_1 (Hz)	f_3 (Hz)	GAIN AT f_3 (dB)-A2	f_5 (Hz)	GAIN AT f_5 (dB)-A3	f_7 (Hz)	GAIN AT f_7 (dB)-A4	f_9 (Hz)	GAIN AT f_9 (dB)-A5
4th Order Butterworth Bandpass Filter Normalized to its Center Frequency, f_{oBP} = 1, and −3dB Bandwidth (BW)																			
1	1	0.693	1.442			0.500	2.000	1.5		2.28	0.500	0.414	−12.3						
1	2	0.836	1.195			0.781	1.281	2.9		2.07	0.781	0.618	−12.3						
1	3	0.885	1.125			0.847	1.180	4.3		2.07	0.847	0.721	−12.3						
1	5	0.932	1.073			0.905	1.105	7.1		2.04	0.905	0.820	−12.3						
1	10	0.965	1.036			0.951	1.051	14.2		2.03	0.951	0.905	−12.3						
1	20	0.982	1.018			0.975	1.025	28.3		2.03	0.975	0.951	−12.3						
6th Order Butterworth Bandpass Filter Normalized to its Center Frequency, f_{oBP} = 1, and −3dB Bandwidth (BW)																			
1	1	0.650	1.539	1.000		0.500	2.000	2.2	1.0	4.79	0.500	0.414	−18.2	0.303	−19.1	0.236	−24.0	0.193	−28.0
1	2	0.805	1.242	1.000		0.781	1.281	4.1	2.0	4.18	0.781	0.618	−18.2	0.500	−19.1	0.414	−24.0	0.351	−28.0
1	3	0.866	1.155	1.000		0.847	1.180	6.1	3.0	4.07	0.847	0.721	−18.2	0.618	−19.1	0.535	−24.0	0.469	−28.0
1	5	0.917	1.091	1.000		0.905	1.105	10.0	5.0	4.03	0.905	0.820	−18.2	0.744	−19.1	0.677	−24.0	0.618	−28.0
1	10	0.958	1.044	1.000		0.951	1.051	20.0	10.0	4.01	0.951	0.905	−18.2	0.861	−19.1	0.820	−24.0	0.781	−28.0
1	20	0.979	1.022	1.000		0.975	1.025	40.0	20.0	4.00	0.975	0.951	−18.2	0.928	−19.1	0.905	−24.0	0.883	−28.0
8th Order Butterworth Bandpass Filter Normalized to its Center Frequency, f_{oBP} = 1, and −3dB Bandwidth (BW)																			
									Q3 = Q4										
1	1	0.809	1.237	0.636	1.574	0.500	2.000	1.1	2.9	10.14	0.500	0.414	−24.0	0.303	−38.0	0.236	−48.1	0.193	−55.8
1	2	0.907	1.103	0.795	1.259	0.781	1.281	2.2	5.4	8.48	0.781	0.618	−24.0	0.500	−38.0	0.414	−48.1	0.351	−55.8
1	3	0.938	1.066	0.858	1.166	0.847	1.180	3.3	7.9	8.15	0.847	0.721	−24.0	0.618	−38.0	0.535	−48.1	0.469	−55.8
1	5	0.962	1.039	0.912	1.097	0.905	1.105	5.4	13.1	8.05	0.905	0.820	−24.0	0.744	−38.0	0.677	−48.1	0.618	−55.8
1	10	0.981	1.019	0.955	1.047	0.951	1.051	10.8	26.2	8.00	0.951	0.905	−24.0	0.861	−38.0	0.820	−48.1	0.781	−55.8
1	20	0.990	1.010	0.977	1.023	0.975	1.025	21.6	52.3	8.00	0.975	0.951	−24.0	0.928	−38.0	0.905	−48.1	0.883	−55.8

Figure 7-1. Butterworth bandpass filters normalized to foBP = 1. (Linear Technology, Application Note 27A, p. 9)

Tabular Design of Bandpass Filters

f_{0BP} (Hz)	$f_{0BP}/BW_1{}^*$ (Hz)	f_{01} (Hz)	f_{02} (Hz)	$f_{0BP}/BW_2{}^{**}$ (Hz)	f_{-3dB} (Hz)	f_{-3dB} (Hz)	Q1=Q2	K	f_1 (Hz)	f_3 (Hz)	GAIN AT f_3(dB)-A2	f_5 (Hz)	GAIN AT f_5(dB)-A3	f_7 (Hz)	GAIN AT f_7(dB)-A4	f_9 (Hz)	GAIN AT f_9(dB)-A5
Passband Ripple, $A_{MAX} = 0.1$dB																	
1	1	0.488	2.050	0.52	0.423	2.364	1.1	3.81	0.500	0.414	−3.2	0.303	−8.7	0.236	−13.6	0.193	−17.4
1	2	0.703	1.422	1.03	0.626	1.597	1.8	2.66	0.781	0.618	−3.2	0.500	−8.7	0.414	−13.6	0.351	−17.4
1	3	0.793	1.261	1.54	0.727	1.375	2.6	2.48	0.847	0.721	−3.2	0.618	−8.7	0.535	−13.6	0.469	−17.4
1	5	0.871	1.148	2.58	0.825	1.213	4.3	2.38	0.905	0.820	−3.2	0.744	−8.7	0.677	−13.6	0.618	−17.4
1	10	0.933	1.071	5.15	0.908	1.102	8.5	2.38	0.951	0.905	−3.2	0.861	−8.7	0.820	−13.6	0.781	−17.4
1	20	0.966	1.035	10.31	0.953	1.050	16.9	2.37	0.975	0.951	−3.2	0.928	−8.7	0.905	−13.6	0.883	−17.4
Passband Ripple, $A_{MAX} = 0.5$dB																	
1	1	0.602	1.660	0.72	0.523	1.912	1.6	3.80	0.500	0.414	−7.9	0.303	−15.0	0.236	−20.2	0.193	−24.1
1	2	0.777	1.287	1.44	0.711	1.406	2.9	3.17	0.781	0.618	−7.9	0.500	−15.0	0.414	−20.2	0.351	−24.1
1	3	0.845	1.182	2.16	0.795	1.258	4.3	3.07	0.847	0.721	−7.9	0.618	−15.0	0.535	−20.2	0.469	−24.1
1	5	0.904	1.106	3.60	0.871	1.149	7.1	3.03	0.905	0.820	−7.9	0.744	−15.0	0.677	−20.2	0.618	−24.1
1	10	0.951	1.051	7.19	0.933	1.072	14.1	2.98	0.951	0.905	−7.9	0.861	−15.0	0.820	−20.2	0.781	−24.1
1	20	0.975	1.025	14.49	0.966	1.035	28.1	2.97	0.975	0.951	−7.9	0.928	−15.0	0.905	−20.2	0.883	−24.1
Passband Ripple, $A_{MAX} = 1.0$dB																	
1	1	0.639	1.564	0.82	0.562	1.779	2.0	4.42	0.500	0.414	−10.3	0.303	−17.7	0.236	−23.0	0.193	−27.0
1	2	0.799	1.251	1.64	0.741	1.349	3.7	3.85	0.781	0.618	−10.3	0.500	−17.7	0.414	−23.0	0.351	−27.0
1	3	0.861	1.161	2.47	0.818	1.223	5.5	3.76	0.847	0.721	−10.3	0.618	−17.7	0.535	−23.0	0.469	−27.0
1	5	0.914	1.094	4.12	0.886	1.129	9.2	3.71	0.905	0.820	−10.3	0.744	−17.7	0.677	−23.0	0.618	−27.0
1	10	0.956	1.046	8.20	0.941	1.063	18.2	3.70	0.951	0.905	−10.3	0.861	−17.7	0.820	−23.0	0.781	−27.0
1	20	0.978	1.022	16.39	0.970	1.031	36.5	3.63	0.975	0.951	−10.3	0.928	−17.7	0.905	−23.0	0.883	−27.0
Passband Ripple, $A_{MAX} = 2.0$dB																	
1	1	0.668	1.496	0.93	0.598	1.672	2.7	6.00	0.500	0.414	−12.7	0.303	−20.3	0.236	−25.5	0.193	−29.5
1	2	0.816	1.225	1.86	0.767	1.304	5.1	5.30	0.781	0.618	−12.7	0.500	−20.3	0.414	−25.5	0.351	−29.5
1	3	0.873	1.145	2.79	0.837	1.195	7.5	5.22	0.847	0.721	−12.7	0.618	−20.3	0.535	−25.5	0.469	−29.5
1	5	0.922	1.085	4.65	0.898	1.113	12.5	5.13	0.905	0.820	−12.7	0.744	−20.3	0.677	−25.5	0.618	−29.5
1	10	0.960	1.041	9.35	0.948	1.055	24.9	5.13	0.951	0.905	−12.7	0.861	−20.3	0.820	−25.5	0.781	−29.5
1	20	0.980	1.021	18.87	0.974	1.027	49.8	5.07	0.975	0.951	−12.7	0.928	−20.3	0.905	−25.5	0.883	−29.5

*f_{0BP}/BW_1 - This is the ratio of the bandpass filter center frequency to the **ripple bandwidth** of the filter.
f_{0BP}/BW_2 - This is the ratio of the bandpass filter center frequency to the **−3dB filter bandwidth.

Figure 7-2. 4th-order Chebyshev bandpass filter normalized to its center frequency fo$BP = 1$. (Linear Technology, Application Note 27A, p. 10)

116 SIMPLIFIED DESIGN OF FILTER CIRCUITS

f_{oBP} (Hz)	$f_{oBP}BW_1$* (Hz)	f_{o1} (Hz)	f_{o2} (Hz)	f_{o3} (Hz)	$f_{oBP}BW_2$** (Hz)	f_{-3dB} (Hz)	f_{-3dB} (Hz)	Q1 = Q2	Q = 3	K	f_1 (Hz)	f_3 (Hz)	GAIN AT f_3 (dB)-A2	f_5 (Hz)	GAIN AT f_5 (dB)-A3	f_7 (Hz)	GAIN AT f_7 (dB)-A4	f_9 (Hz)	GAIN AT f_9 (dB)-A5
Passband Ripple, A_{MAX} = 0.1dB																			
1	1	0.558	1.791	1.000	0.72	0.523	1.912	2.4	1.0	9.9	0.500	0.414	−12.2	0.303	−23.6	0.236	−31.4	0.193	−37.3
1	2	0.741	1.349	1.000	1.44	0.711	1.406	4.3	2.1	7.9	0.781	0.618	−12.2	0.500	−23.6	0.414	−31.4	0.351	−37.3
1	3	0.818	1.222	1.000	2.16	0.795	1.258	6.3	3.1	7.5	0.847	0.721	−12.2	0.618	−23.6	0.535	−31.4	0.469	−37.3
1	5	0.886	1.128	1.000	3.60	0.871	1.149	10.4	5.2	7.4	0.905	0.820	−12.2	0.744	−23.6	0.677	−31.4	0.618	−37.3
1	10	0.941	1.062	1.000	7.19	0.933	1.072	20.6	10.3	7.3	0.951	0.905	−12.2	0.861	−23.6	0.820	−31.4	0.781	−37.3
1	20	0.970	1.030	1.000	14.49	0.966	1.035	41.3	20.6	7.3	0.975	0.951	−12.2	0.928	−23.6	0.905	−31.4	0.883	−37.3
Passband Ripple, A_{MAX} = 0.5dB																			
1	1	0.609	1.641	1.000	0.86	0.574	1.741	3.6	1.6	14.8	0.500	0.414	−19.2	0.303	−30.8	0.236	−38.6	0.193	−44.5
1	2	0.776	1.288	1.000	1.72	0.750	1.333	6.6	3.2	12.5	0.781	0.618	−19.2	0.500	−30.8	0.414	−38.6	0.351	−44.5
1	3	0.844	1.185	1.000	2.57	0.824	1.213	9.7	4.8	12.0	0.847	0.721	−19.2	0.618	−30.8	0.535	−38.6	0.469	−44.5
1	5	0.903	1.107	1.000	4.29	0.890	1.123	16.1	8.0	11.8	0.905	0.820	−19.2	0.744	−30.8	0.677	−38.6	0.618	−44.5
1	10	0.950	1.052	1.000	8.55	0.943	1.060	32.0	16.0	11.8	0.951	0.905	−19.2	0.861	−30.8	0.820	−38.6	0.781	−44.5
1	20	0.975	1.026	1.000	16.95	0.971	1.030	63.8	32.0	11.4	0.975	0.951	−19.2	0.928	−30.8	0.905	−38.6	0.883	−44.5
Passband Ripple, A_{MAX} = 1.0dB																			
1	1	0.626	1.598	1.000	0.91	0.593	1.687	4.5	2.0	20.1	0.500	0.414	−22.5	0.303	−34.0	0.236	−41.9	0.193	−47.8
1	2	0.787	1.271	1.000	1.83	0.763	1.310	8.3	4.1	17.1	0.781	0.618	−22.5	0.500	−34.0	0.414	−41.9	0.351	−47.8
1	3	0.852	1.174	1.000	2.74	0.834	1.199	12.3	6.1	16.7	0.847	0.721	−22.5	0.618	−34.0	0.535	−41.9	0.469	−47.8
1	5	0.908	1.101	1.000	4.59	0.897	1.115	20.3	10.1	16.4	0.905	0.820	−22.5	0.744	−34.0	0.677	−41.9	0.618	−47.8
1	10	0.953	1.050	1.000	9.17	0.947	1.056	40.5	20.2	16.4	0.951	0.905	−22.5	0.861	−34.0	0.820	−41.9	0.781	−47.8
1	20	0.976	1.024	1.000	18.18	0.973	1.028	81.0	40.5	16.4	0.975	0.951	−22.5	0.928	−34.0	0.905	−41.9	0.883	−47.8
Passband Ripple, A_{MAX} = 2.0dB																			
1	1	0.639	1.565	1.000	0.97	0.609	1.642	6.0	2.7	31.7	0.500	0.414	−26.0	0.303	−37.5	0.236	−45.4	0.193	−51.3
1	2	0.795	1.257	1.000	1.94	0.775	1.291	11.1	5.4	27.4	0.781	0.618	−26.0	0.500	−37.5	0.414	−45.4	0.351	−51.3
1	3	0.858	1.165	1.000	2.91	0.843	1.187	16.5	8.1	26.7	0.847	0.721	−26.0	0.618	−37.5	0.535	−45.4	0.469	−51.3
1	5	0.912	1.096	1.000	4.83	0.902	1.109	27.2	13.6	26.2	0.905	0.820	−26.0	0.744	−37.5	0.677	−45.4	0.618	−51.3
1	10	0.955	1.047	1.000	9.71	0.950	1.053	54.3	27.1	26.0	0.951	0.905	−26.0	0.861	−37.5	0.820	−45.4	0.781	−51.3
1	20	0.977	1.023	1.000	19.61	0.975	1.026	108.5	54.2	26.0	0.975	0.951	−26.0	0.928	−37.5	0.905	−45.4	0.883	−51.3

*$f_{oBP}BW_1$ – This is the ratio of the bandpass filter center frequency to the **ripple bandwidth** of the filter.
$f_{oBP}BW_2$ – This is the ratio of the bandpass filter center frequency to the **−3dB filter bandwidth.

Figure 7-3. 6th-order Chebyshev bandpass filter normalized to its center frequency foBP = 1. (Linear Technology, Application Note 27A, p. 11)

Tabular Design of Bandpass Filters   117

f_{oBP} (Hz)	f_{oBP}/BW_1*	f_{o1} (Hz)	f_{o2} (Hz)	f_{o3} (Hz)	f_{o4} (Hz)	f_{oBP}/BW_2**	f_{-3dB} (Hz)	f_{-3dB} (Hz)	Q1 = Q2	Q3 = Q4	K	f_1 (Hz)	f_3 (Hz)	GAIN AT f_3 (dB)-A2	f_5 (Hz)	GAIN AT f_5 (dB)-A3	f_7 (Hz)	GAIN AT f_7 (dB)-A4	f_9 (Hz)	GAIN AT f_9 (dB)-A5
Passband Ripple, A_{MAX} = 0.1dB																			
1	1	0.785	1.274	0.584	1.713	0.82	0.563	1.776	1.6	4.4	40.6	0.500	0.414	−23.4	0.303	−38.8	0.236	−49.3	0.193	−57.1
1	2	0.889	1.125	0.757	1.320	1.65	0.742	1.348	3.2	7.9	32.1	0.781	0.618	−23.4	0.500	−38.8	0.414	−49.3	0.351	−57.1
1	3	0.925	1.081	0.830	1.204	2.48	0.818	1.222	4.7	11.6	30.5	0.847	0.721	−23.4	0.618	−38.8	0.535	−49.3	0.469	−57.1
1	5	0.954	1.048	0.894	1.118	4.12	0.886	1.129	7.9	19.1	29.9	0.905	0.820	−23.4	0.744	−38.8	0.677	−49.3	0.618	−57.1
1	10	0.977	1.023	0.945	1.058	8.20	0.941	1.063	15.7	37.9	29.8	0.951	0.905	−23.4	0.861	−38.8	0.820	−49.3	0.781	−57.1
1	20	0.988	1.012	0.972	1.028	16.39	0.970	1.031	31.4	75.7	29.8	0.975	0.951	−23.4	0.928	−38.8	0.905	−49.3	0.883	−57.1
Passband Ripple, A_{MAX} = 0.5dB																			
1	1	0.808	1.238	0.613	1.632	0.91	0.593	1.686	2.4	6.4	90.1	0.500	0.414	−30.2	0.303	−45.5	0.236	−56.0	0.193	−63.9
1	2	0.900	1.111	0.777	1.286	1.83	0.763	1.310	4.8	11.8	74.3	0.781	0.618	−30.2	0.500	−45.5	0.414	−56.0	0.351	−63.9
1	3	0.932	1.073	0.845	1.183	2.74	0.834	1.199	7.1	17.4	71.5	0.847	0.721	−30.2	0.618	−45.5	0.535	−56.0	0.469	−63.9
1	5	0.959	1.043	0.903	1.107	4.59	0.897	1.115	11.8	28.7	70.0	0.905	0.820	−30.2	0.744	−45.5	0.677	−56.0	0.618	−63.9
1	10	0.979	1.021	0.950	1.052	9.17	0.947	1.056	23.6	57.1	70.0	0.951	0.905	−30.2	0.861	−45.5	0.820	−56.0	0.781	−63.9
1	20	0.989	1.010	0.975	1.026	18.18	0.973	1.028	47.2	114.0	70.0	0.975	0.951	−30.2	0.928	−45.5	0.905	−56.0	0.883	−63.9
Passband Ripple, A_{MAX} = 1.0dB																			
1	1	0.814	1.228	0.622	1.607	0.95	0.604	1.656	3.0	8.0	162.8	0.500	0.414	−32.9	0.303	−48.3	0.236	−58.8	0.193	−66.6
1	2	0.903	1.107	0.784	1.275	1.90	0.771	1.297	6.0	14.8	133.2	0.781	0.618	−32.9	0.500	−48.3	0.414	−58.8	0.351	−66.6
1	3	0.934	1.070	0.850	1.177	2.85	0.840	1.191	8.9	21.8	128.1	0.847	0.721	−32.9	0.618	−48.3	0.535	−58.8	0.469	−66.6
1	5	0.960	1.041	0.906	1.103	4.74	0.900	1.111	14.9	36.0	127.7	0.905	0.820	−32.9	0.744	−48.3	0.677	−58.8	0.618	−66.6
1	10	0.980	1.020	0.952	1.050	9.52	0.949	1.054	29.7	71.7	124.0	0.951	0.905	−32.9	0.861	−48.3	0.820	−58.8	0.781	−66.6
1	20	0.990	1.010	0.976	1.025	18.87	0.974	1.027	59.4	143.0	120.0	0.975	0.951	−32.9	0.928	−48.3	0.905	−58.8	0.883	−66.6
Passband Ripple, A_{MAX} = 2.0dB																			
1	1	0.820	1.220	0.629	1.589	0.98	0.613	1.631	4.0	10.6	374.8	0.500	0.414	−35.4	0.303	−50.8	0.236	−61.3	0.193	−69.2
1	2	0.905	1.104	0.789	1.266	1.96	0.777	1.287	7.9	19.6	312.6	0.781	0.618	−35.4	0.500	−50.8	0.414	−61.3	0.351	−69.2
1	3	0.936	1.068	0.853	1.172	2.95	0.845	1.184	11.9	29.0	302.0	0.847	0.721	−35.4	0.618	−50.8	0.535	−61.3	0.469	−69.2
1	5	0.961	1.040	0.909	1.100	4.90	0.903	1.107	19.7	47.9	302.0	0.905	0.820	−35.4	0.744	−50.8	0.677	−61.3	0.618	−69.2
1	10	0.980	1.020	0.953	1.049	9.80	0.950	1.052	39.5	95.4	302.0	0.951	0.905	−35.4	0.861	−50.8	0.820	−61.3	0.781	−69.2
1	20	0.990	1.010	0.976	1.024	19.61	0.975	1.026	79.0	190.0	302.0	0.975	0.951	−35.4	0.928	−50.8	0.905	−61.3	0.883	−69.2

*f_{oBP}/BW_1 – This is the ratio of the bandpass filter center frequency to the **ripple bandwidth** of the filter.
f_{oBP}/BW_2 – This is the ratio of the bandpass filter center frequency to the **−3dB filter bandwidth.

Figure 7–4. 8th-order Chebyshev bandpass filter normalized to its center frequency foBP = 1. (Linear Technology, Application Note 27A, p. 12)

Figure 7-5.
Generalized Butterworth bandpass response. (Linear Technology, Application Note 27A, p. 8)

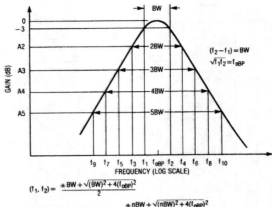

$$(f_1, f_2) = \frac{\pm BW + \sqrt{(BW)^2 + 4(f_{oBP})^2}}{2}$$

MORE GENERALLY $(f_x, f_{x+1}) = \dfrac{\pm nBW + \sqrt{(nBW)^2 + 4(f_{oBP})^2}}{2}$

(VALID FOR ANY f_x, f_{x+1} PAIR, ANY BW)

Figure 7-6.
Generalized Chebyshev bandpass response with 2-dB passband ripple. (Linear Technology, Application Note 27A, p. 8)

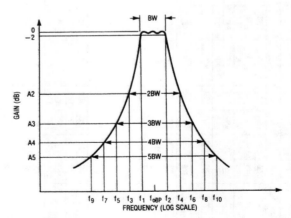

$\sqrt{f_4 f_3} = f_{oBP}$

$(f_4, f_3) = \dfrac{\pm 2BW + \sqrt{(2BW)^2 + 4(f_{oBP})^2}}{2}$

FOR ANY (f_x, f_{x+1}) PAIR AND
ANY CORRESPONDING BANDWIDTH
(2BW, 3BW, ETC.)

FOR EXAMPLE:

$(f_6, f_5) = \dfrac{\pm 3BW + \sqrt{(3BW)^2 + 4(f_{oBP})^2}}{2}$

symmetrical about their center frequencies, foBP. Any frequency f3, as shown in Fig. 7–5, has its geometrical counterpart f4 such that:

$$f4 = (foBP^2) / f3$$

The table of Fig. 7–1 also shows the attenuation of the frequencies f3, f5, f7, and f9 that correspond to bandwidths 2, 3, 4, and 5 times the passband (Fig. 7–5). These values allow the user to get a good estimate of filter selectivity.

An important approximation can be made for not only the Butterworth filters in Fig. 7–1, but also for the Chebyshev filters of Figs. 7–2, 7–3, and 7–4. By treating Fig. 7–5 (or Fig. 7–6) as a generalized bandpass filter, the two corner frequencies f2 and f1 can be seen as nearly arithmetically symmetrical with respect to foBP provided that:

$$\frac{foBP}{BW} \gg \frac{1}{2}, \text{ where } BW = f2 - f1$$

Under this condition, for either Butterworth or Chebyshev bandpass filters:

$$foBP = \frac{f3 - f4}{2} + f3$$

$$foBP = \frac{f5 - f6}{2} + f5$$

and so on. This is true for any bandwidth, BW, and any set of frequencies. The tables can now be arithmetically scaled as shown.

7.2 Gain and Phase Relationships of the IC Filters

Before we get into actual design, let use review the characteristics of the Linear Technology filters involved. The bandpass output of each 2nd-order filter section of the LTC1059, 60, 61, and 64 closely approximates the gain and phase response of a textbook filter. This is shown in Figs. 7–7a and 7–7b.

7.2.1 Bandpass Gain versus Q

Figure 7–8 shows the bandpass gain, G, for various values of Q, and is very useful for estimating the filter attenuation when several identical 2nd-order bandpass filters are cascaded. High values of Q make a filter more selective, and at the same time, more noisy and more difficult to realize. Qs in excess of 100 can be realized with universal switched-capacitor filters (such as the LTC1059, 60, 61, 64), and still maintain low center frequency and Q drift. However, for system considerations, this is usually not practical. The problem is one of phase shift.

120 SIMPLIFIED DESIGN OF FILTER CIRCUITS

$$G = \frac{(H_{0BP}) \times (ff_0)/Q}{[(f_0^2 - f^2)^2 + (ff_0/Q)^2]^{1/2}}$$

G = filter gain in Volts/V

f_0 = the filter's center frequency

Q = the quality coefficient of the filter

H_{0BP} = the maximum voltage gain of the filter occurring at f_0

$\dfrac{f_0}{Q}$ = the −3dB bandwidth of the filter

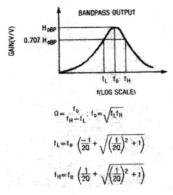

Figure 7-7.
Gain and phase relationships of bandpass filters. (Linear Technology, Application Note 2)

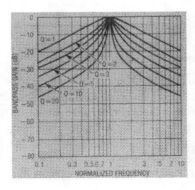

Figure 7-8.
Bandpass gain as a function of Q. (Linear Technology, Application Note 27A, p. 6)

7.2.2 Phase Shift in Filters

Figure 7–9 shows the relationship of phase shift, frequency, and Q of a 2nd-order bandpass filter. The phase shift at fo is 0°. If the filter is inverting, the phase shift is –180°. (All bandpass outputs of the LTC1059, 60, 61, and 64 filters are inverting.) The phase shift, especially near fo, depends on the value of Q. The phase shift at a given frequency varies from IC to IC because of the fo tolerance. This is especially true for high-Q designs and at frequencies near fo.

For example, an LTC1059A 2nd-order filter has a guaranteed initial center-frequency tolerance of ±0.3 percent. The ideal phase shift at the ideal fo should be –180°. With a Q of 20, and without trimming, the worst-case phase shift at the ideal fo will be –180° ±6.8°. With a Q of 5, the phase-shift tolerance becomes –180° ±1.7°.

Obviously, the relationship of Q to phase shift is an important consideration when bandpass filters are used in multichannel systems where phase matching is required. By way of comparison, a state-variable active bandpass filter using 1 percent resistors and 1 percent capacitors might have center-frequency variation of ±2 percent, resulting in a phase variation of ±38.8° for a Q of 20 and ±11.4° for a Q of 5.

7.2.3 Constant Q versus Constant Bandwidth

The bandpass output of the universal filters is ideally constant Q. For example, a 2nd-order bandpass filter operating in mode 1 with a 100-kHz clock ideally has a 1-kHz or 2-kHz center frequency and a –3-dB bandwidth equal to fo/Q. When the clock frequency varies, the center frequency and bandwidth will vary at the same rate. In a constant-bandwidth filter, when the center frequency varies, the Q varies accordingly to maintain a constant fo/Q ratio.

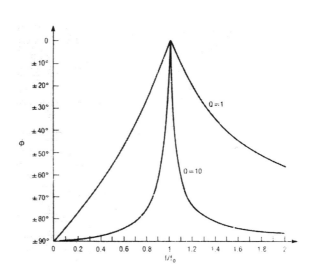

Figure 7–9. Relationship of phase shift, frequency, and Q of a 2nd-order bandpass filter. (Linear Technology, Application Note 27A, p. 7)

7.3 Cascading Non-Identical IC Filters

Now let us use the tables to design a 4th-order 2-kHz Butterworth bandpass filter with a –3-dB bandwidth equal to 200 Hz. This provides a response similar to that of Fig. 7–10.

Because we want Butterworth response, we use the table of Fig. 7–1. First find the center-frequency to bandwidth ratio (foBP/BW). In our example, this ratio of 10 (2 kHz/200 Hz). Now look up 10 under the 4th-order section to find fo1 and fo2. These are 0.965 and 1.036 (both normalized to an foBP of 1). To find our desired actual center frequencies, we must multiply by foBP = 2 kHz to get fo1 = 1.930 kHz and fo2 = 2.072 kHz.

Next find the Qs (Q1 = Q2 = 14.2) using the table. Also available from the table is K, which is the product of each individual bandpass gain HoBP. In effect, the value of K is the gain required to make the gain, H, of the overall filter equal to 1 at foBP.

We have now established the parameters needed for simplified design using the tabular method. These parameters are: foBP = 2 kHz, fo1 (the fo of the first section) = 1.93 kHz, fo2 (the fo of the second section) = 2.072 kHz, Q1 = Q2 = 14.2, and K = 2.03.

There are several ways such a filter could be implemented. For example, we could use state-variable filters (either switched-capacitor or active) as shown in Fig. 7–11. Instead, we will use two non-identical filters in cascade as shown in Fig. 7–12. Note that the cascaded filters are two 2nd-order sections of the Linear Technology LTC family (1 LTC1060 and 2/3 of an LTC1061, or 1/2 of an LTC1064).

The next step is to relate the resistors in Fig. 7–12 to the equations of Fig. 7–12; that is, R1 of the equations is R11 in the first (or input) section and R12 in the second (or output) section. R2 = R21 and R22, R3 = R31 and R32, R4 = R41 and R42.

The equations of Fig. 7–12 show that fo is set by the clock frequency, and the values of R2 and R4. Arbitrarily set the clock at 100 kHz. (In the LTC family of switched-capacitor ICs, it is necessary to connect the 50/100/Hold pin to V+ for a clock of 100 kHz. There is usually some similar connection for other universal switched-capacitor ICs.)

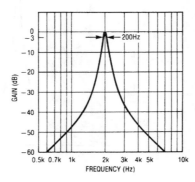

Figure 7–10.
4th-order Butterworth bandpass filter response for an foBP of 2 kHz. (Linear Technology, Application Note 27A, p. 1)

Tabular Design of Bandpass Filters 123

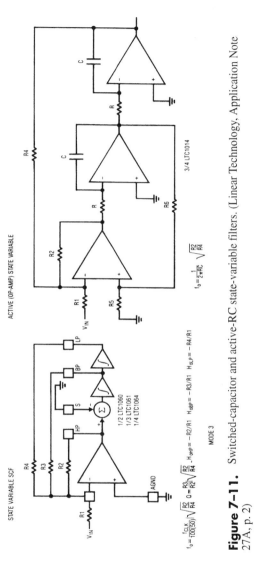

Figure 7-11. Switched-capacitor and active-RC state-variable filters. (Linear Technology, Application Note 27A, p. 2)

124 SIMPLIFIED DESIGN OF FILTER CIRCUITS

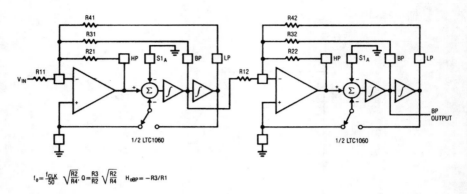

$f_o = \frac{f_{CLK}}{50} \sqrt{\frac{R2}{R4}}; \quad Q = \frac{R3}{R2}\sqrt{\frac{R2}{R4}}; \quad H_{oBP} = -R3/R1$

Figure 7-12. Two non-identical 2nd-order sections cascaded to form a 4th-order bandpass filter. (Linear Technology, Application Note 27A, p. 3)

The K factor given in the tables is the product of HoBP for both sections. For simplification, set HoBP for the input section at 1. Then the HoBP for the output section is 1 times 2.03 (to get a K of 2.03).

We now have the values needed for the equations in Fig. 7–12. These values are:

Input Section	*Output Section*
fo1 = 1.93 kHz (or 0.965)	fo2 = 2.072 kHz (or 1.036)
Q1 = 14.2	Q2 = 14.2
HoBP1 = 1	HoBP2 = 2.03

Now let us find the resistor values for the input section using the equations of Fig. 7–12. For convenience, set R21(R2) at 10 k. This will simplify both the fo and Q equations. fCLK/50 can be reduced to 2,000 because we are using a 100 kHz clock (for both sections). We also know that R3 = R1 (for the input section) because HoBP1 = 1, and is the ratio of R3/R1.

For the input section, fo(1.93kHz) = 2000√10k/R41. This can be rearranged as √10k/R41 = 1930/2000 = 0.965. By removing the square root, we find that 10k/R41 = 0.931225. Further rearrangement produces R41 = 10000/0.931225 = 10738. The nearest 1 percent standard resistor is 10.7 k for R41.

For the input section, Q(14.2) = R3/10k√10k/10.7k. This can be reduced to Q(14.2) = R3/10k√0.934579. By removing the square root, we find that Q(14.2) = R3/10k 0.966. This can be rearranged to R3/10k = 14.2/0.966 = 14.69. Further rearrangement produces R3 = 14.69 × 10000 = 146900. The nearest 1 percent standard resistor is 147 k. With R3 at 147 k, R1 is also 147 k because HoBP1 is 1.

As a result of these calculations, the input-section resistance values are R11 = 147 k, R21 = 10 k, R31 = 147 k, and R41 = 10.7 k.

The resistor values for the second or output section can be found in the same manner. However, remember that fo2 is 2.072 kHz, and HoBP2 is 2.03, so R3 must be 2.03 times the value of R1. The Linear Technology literature lists the resistor values for the second or output section as R21 = 71.5 k, R22 = 10.7 k, R32 = 147 k, and R42 = 10k.

7.4 Cascading Identical IC Filters

Now let us design a 2nd-order bandpass filter that will pass 150 Hz without attenuation, but will attenuate 60-Hz signals by 50 dB. This results in a response similar to that of Fig. 7–13b. Such a response could be obtained with a single 2nd-order bandpass filter, but the Q would be very high! Instead, we will use two identical 2nd-order filters connected in cascade.

We do not need the tables for this design, but will use two identical sections of the Linear Technology LTC family (both sections of an LTC1060). We describe two separate approaches, one using mode 1 and the other using mode 2. (If you are not familiar with mode 1 and mode 2, please review Chapter 2.)

Before we get into the designs of the specific circuits, let us review the relationship of filter Q, bandpass, and attenuation.

For a 2nd-order bandpass filter, find the Q as follows:

$$Q = \frac{\sqrt{1 - G^2}}{G} \frac{f/f_o}{[1 - (f/f_o)^2]}$$

where Q is the required filter quality factor (sharpness), f is the frequency where the filter should have gain, G, expressed in volts/V.

f_o is the filter center frequency. Unity gain is assumed at f_o.

Using this 2nd-order bandpass filter equation, calculate the required Q to get the desired bandpass and attenuation as follows:

$$Q = \frac{\sqrt{1 - (3.162 \times 10^{-3})^2}}{3.162 \times 10^{-3}} \frac{60/150}{[1 - (60/150)^2]} = 150.7$$

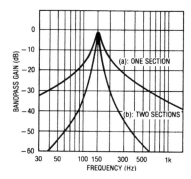

Figure 7–13. Comparison of response for one section versus two sections in cascade. (Linear Technology, Application Note 27A, p. 5)

126 SIMPLIFIED DESIGN OF FILTER CIRCUITS

This very high Q will require a –3dB bandwidth of 1 Hz.

Although universal switched-capacitor filters such as the LTC1060 can achieve high Qs, their guaranteed center-frequency accuracy of 0.3 percent is not enough to pass the 150-Hz signal without gain errors. According to the 2nd-order Q equation, the gain at 150 Hz will be 1 ± 26 percent. However, the rejection at 60 Hz will remain at –50 dB. The gain inaccuracy can be corrected by selection (tuning) of resistor R4 when mode 3 operation (Fig. 7–11) is used. Also, if only detection of the signal is sought, the gain inaccuracy could be acceptable.

The high Q problem can be solved by cascading two identical 2nd-order bandpass sections. To get a gain, G, at frequency f, find the required Q of each section as follows:

$$Q = \frac{\sqrt{1-G}}{\sqrt{G}} \frac{f/fo}{[1-(f/fo)^2]}$$

The gain at each bandpass section is assumed as unity.

Calculate the required Q for each 2nd-order section as follows:

$$Q = \frac{\sqrt{1-3.162\times10^{-3}}}{\sqrt{3.162\times10^{-3}}} \frac{60/150}{[1-(60/150)^2]} = 8.5$$

With two identical 2nd-order section, each with a potential error in center frequency, fo, of ±0.3 percent, the gain error at 150 Hz is 1 ±0.26 percent. If lower-cost (non-A versions of the LTC family) 2nd-order bandpass sections are used with an fo tolerance of ±0.8 percent, the gain error at 150 Hz is 1 ±1.8% percent.

7.5 Cascading Bandpass Filters in Mode 1

Figure 7–14 shows two cascaded 2nd-order bandpass filters (two sections of an LTC1060) connected for mode-1 operation. Again, the resistors must be related to the

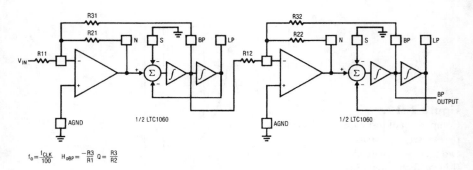

Figure 7–14. LTC1060 connected as a cascaded bandpass filter operating in mode 1. (Linear Technology, Application Note 27A, p. 4)

equations as discussed in Section 7.3. That is, R1 = R11 and R12, R2 = R21 and R22, R3 = R31 and R32, R4 = R41 and R42.

However, unlike the cascaded filters in Section 7.3, both sections (input and output) of the filters in Fig. 7–14 must have the same requirements. That is, fo1 = fo2 = 150 Hz, Q1 = Q2 = 8.5, HoBP1 = HoBP2 = 1.

For our example, using the LTC1060, we let fo1 = fo2 = fCLK/100. This requires a 15-Hz clock. (It also means that we must tie the 50/100/Hold pin to mid-supplies, or to ground for ±5-V supplies.)

Mode 1 is the fastest operating mode for our switched-capacitor filters, and should result in a response similar to that of Fig. 7–13b *when both sections are cascaded.* Each section should produce a response similar to that of Fig. 7–13a (about 25 dB of attenuation at 60 Hz, with unity gain at 150 Hz).

Calculate the resistor values using the equations of Fig. 7–14. Note that these calculations are considerably simpler than those for non-identical sections (Fig. 7–12). However, 1 percent resistors should still be used, and the minimum resistance value should be 20 k (for R21 and R22).

The Linear Technology literature recommends the following resistor values for mode-1 operation as shown in Fig. 7–14 to get a response similar to that of Fig. 7–13b. R11 = R12 = 169k, R21 = R22 = 20k, R31 = R32 = 160k.

7.6 Cascading Bandpass Filters in Mode 2

Figure 7–15 shows two cascaded 2nd-order bandpass filters (again two sections of an LTC1060) connected for mode-2 operation. Again, the resistors must be related to the equations on Fig. 7–15 as discussed, and both sections must be identical.

Mode 2 is used when there is no 15-kHz clock readily available. This allows the input clock frequency to be less than 50:1 or 100:1, but still requires that the

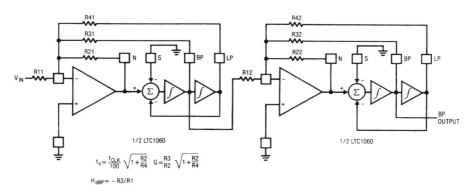

Figure 7–15. LTC1060 connected as a cascaded bandpass filter operating in mode 2. (Linear Technology, Application Note 27A, p. 5)

128 SIMPLIFIED DESIGN OF FILTER CIRCUITS

50/100/Hold pin be properly connected. For example, we could operate the filter from a 14.318-MHz television crystal by dividing the crystal oscillator output by 1000.

The Linear Technology literature recommends the following resistor values for mode-2 operation as shown in Fig. 7–15 to get a response similar to that of Fig. 7–13b: R11 = R12 = 162 k, R21 = R22 = 20 k, R31 = R32 = 162 k, R41 = R42 = 205 k.

7.7 Cascading More Than Two Identical 2nd-Order Sections

If more than two identical 2nd-order bandpass sections are cascaded, calculate the required Q of each section as follows:

$$Q = \frac{\sqrt{1 - G^{2/n}}}{G^{1/n}} \frac{(f/fo)}{[1 - (f/fo)^2]}$$

where Q, G, f, and fo are as previously defined, and n = the number of cascaded 2nd-order sections.

The equivalent of Q of the overall bandpass filter is then:

$$Qequiv = \frac{Q(\text{identical section})}{\sqrt{(2^{1/n})} = 1}$$

Figure 7–16 shows the passband curves for Q = 2 cascaded bandpass sections where n is the number of 2nd-order sections. As a practical matter, cascading two or three sections provides obvious benefits. However, cascading four (or more) sections does not usually improve the Q that much considering the additional hardware involved.

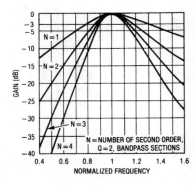

Figure 7–16. Frequency response of n cascaded identical 2nd-order bandpass sections. (Linear Technology, Application Note 27A, p. 6)

CHAPTER **8**

Practical Considerations for Switched-Capacitor Filters

This chapter is devoted to the practical side of switched-capacitor filters (SCFs), stressing both their capabilities and limitations. The discussion revolves around the Linear Technology IC filters, but also applies to all present-day SCFs in IC form. We start with a comparison of SCFs and active RC filters. (Chapter 9 describes the simplified design approach for active RC filters.)

8.1 SCFs versus Active RC Filters

Figures 8–1 and 8–2 show a comparison between SCFs and active RC filters. In both cases, the building block is the integrator. When the integrator is implemented with resistors, capacitors, and op amps (as described in Chapter 9), the design is expensive and sensitive to component tolerances. The SCF integrator (Fig. 8–2) eliminates the resistors and replaces them with switched capacitors.

The SCF integrator depends on capacitor-value ratios, and not on absolute values. This provides very good accuracy in setting center frequencies and Q values. Also, the SCF approach allows the effective resistor value to be varied by the clock that operates the switches. In turn, this allows the resistor (and thus the filter center frequencies) to be varied over a wide range (typically 10,000:1 or more).

Because IC technology can implement capacitor ratios much more accurately than resistor ratios, the SCF can provide filters with inherent accuracy and repeatability. Active RC filters are limited by resistor and capacitor tolerances (and to a secondary extent, the accuracy and bandwidth of the op amps) and usually require trimming. However, SCFs do have disadvantages compared to active RC filters. One special problem with SCFs is clock feedthrough, where the clock signal appears at the filter output. This has the same effect on the external circuits as noise.

Figure 8-1.
Active RC inverging integrator. (Linear Technology, Application Note 40, p. 1)

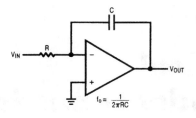

Figure 8-2.
Inverting SCF integrator. (Linear Technology, Application Note 40, p. 1)

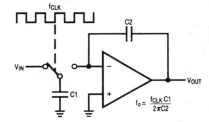

8.2 Circuit Board Layout Problems

Figure 8–3 shows the circuit connections for a Linear Technology LTC1064-1 SCF and an LT1007CN buffer amplifier used for breadboard tests. This circuit was used to compare two breadboard techniques. One breadboard has no bypass capacitors, no ground plane, and the clock was connected with wires instead of coax. The tests run on this "poor breadboard" were made without the buffer amplifier. The recommended breadboard was cooper-clad to provide a ground plane, all leads were kept as short as possible, and the SCF clock input was made through a shielded cable. Also, the recommended breadboard used single-point grounding throughout. (Single-point grounding, sometimes known as star grounding, is essential when SCFs are used in large systems where boards contain many analog and digital devices.)

Figures 8–4 through 8–7 show a comparison of the breadboards. The following is a summary of the test results.

Figure 8–4 shows the passband response of an 8th-order Elliptic lowpass filter (the LTC1064-1). The ripple in the passband is caused by the breadboard techniques rather than by the filter. (The top trace is made with the poor breadboard.) In both cases, the clock is 2 MHz, the cutoff is at 20 kHz, and VS is ±8 V. The top trace was offset to increase clarity of the display. The recommended breadboard is a factor of 5 to 10 times better than the poor breadboard.

Figure 8–5 shows the stopband response of the 8th-order Elliptical lowpass filter. Again, the top trace is that of the poor breadboard, the clock is 2 MHz, the cutoff is at 20 kHz, and VS is ±8 V. Note the loss of between 10 dB and 20 dB of attenuation. Also notice the notch, almost 80 dB down from the input signal, which is clearly shown when using the recommended breadboard, and which cannot be seen using the poor breadboard.

Practical Considerations for Switched-Capacitor Filters 131

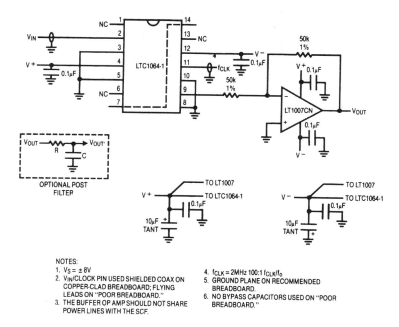

Figure 8-3. C1064-1 SCF and LT1007CN buffer-amp breadboard circuit. (Linear Technology, Application Note 40, p. 3)

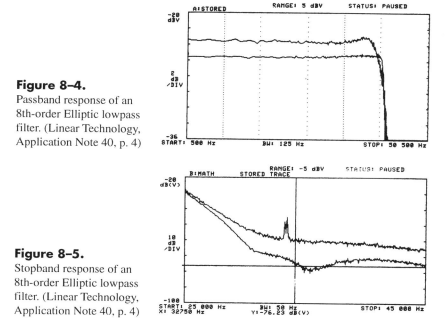

Figure 8-4.
Passband response of an 8th-order Elliptic lowpass filter. (Linear Technology, Application Note 40, p. 4)

Figure 8-5.
Stopband response of an 8th-order Elliptic lowpass filter. (Linear Technology, Application Note 40, p. 4)

132 SIMPLIFIED DESIGN OF FILTER CIRCUITS

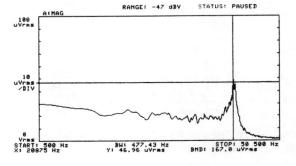

Figure 8-6.
Wideband response with improper breadboard techniques. (Linear Technology, Application Note 40, p. 4)

Figure 8-7.
Wideband response with good breadboard techniques. (Linear Technology, Application Note 40, p. 4)

Figures 8–6 and 8–7 show the noise for the two breadboards. Again, the clock is 2 MHz, cutoff is 20 kHz, and VS is ±8 V. Two plots were necessary because the noise was more than an order of magnitude greater when the poor breadboard (Fig. 8–6) was used. The tests were run using the identical SCF (LTC1064-1CN) which was moved from one breadboard to the other to ensure exactly the same measurement conditions (except for the breadboards).

Not shown graphically, but measured, was the offset voltage of the SCF IC in each breadboard. In the poor breadboard, the offset was 266 mV. Offset for the IC when tested in the recommended breadboard was 40.7 mV (a factor of almost seven times less).

Note that all measurements described in this chapter (with a few noted exceptions) were performed with a good 10X, low-capacitance scope probe (Tektronix P6133) at the output of the buffer amp. The probe ground lead was kept below one inch in length.

8.3 Power Supply Problems

As is the case for any sampled data device (A/D and D/A converters, chopper-stabilized op amps, and sampled-data comparators), the power supplies for SCFs require careful design. Bypassing the supply lines is a particular problem. Poorly chosen or poorly designed power supplies can actually induce noise into the system. Similarly, improper bypassing can impair even the most ideal of supplies. Common problems include noise in the passband of the SCF and spurious A/D output because of high-frequency noise being aliased back into the signal bandwidth of interest.

Figure 8–8 shows an example of typical power-supply problems. The display was made using the poor breadboard, with industry standard +5 V to ±15 V switcher (converter) as the power supply. The supply was unbypassed to better illustrate the potential problems. The display is that of stopband response with a clock of 2 MHz, a 20 kHz cutoff, and a VS of ±7.5 V (the supply output was zenered at ±7.5 V). The filter is still an 8th-order Elliptic lowpass using the LTC1064-1.

Figure 8–8 can be compared directly with the display of Fig. 8–5 to illustrate the noise. The poor breadboard causes the stopband attenuation to be well above where it should be when proper breadboarding techniques are used, but switcher harmonics are also evident. These appear in Fig. 8–8. Some of these peaks are only –45 dB down from the signal of interest in the passband, and could be confused with a legitimate signal.

Figure 8–9 shows the schematic diagram of a good, low-noise converter for SCF use. The supply produces ±7.5 V with 20 µV of noise. (For a detailed discussion of DC-DC converters, and converter noise, read the author's *Simplified Design of Switching Power Supplies,* 1995, Butterworth-Heinemann.)

For SCFs such as the LTC1064, the manufacturer recommends good, low-ESR (equivalent series resistance) bypass capacitors, with a value of 0.1 µF minimum (0.22 µF is better). Always connect the bypass capacitors as close to the power-supply pins of the SCF IC as possible (where have we heard that before?). Always use high-quality capacitors for SCFs (and any other IC where you really care about performance!). There is more on bypass capacitors at the end of this chapter.

Separate digital and analog grounds are recommended. When both ground must be tied together, make the connection as close to supply common as practical. The ground lead of bypass capacitors should go to the analog ground plane.

Bypass any pins on the SCF that are tied to a circuit potential for programming (such as the 50/100/Hold pin). If spikes and/or transients appear on this pin, problems

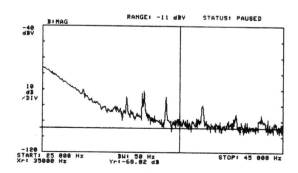

Figure 8–8.
Stopband response with power-supply problems. (Linear Technology, Application Note 40, p. 5)

134 SIMPLIFIED DESIGN OF FILTER CIRCUITS

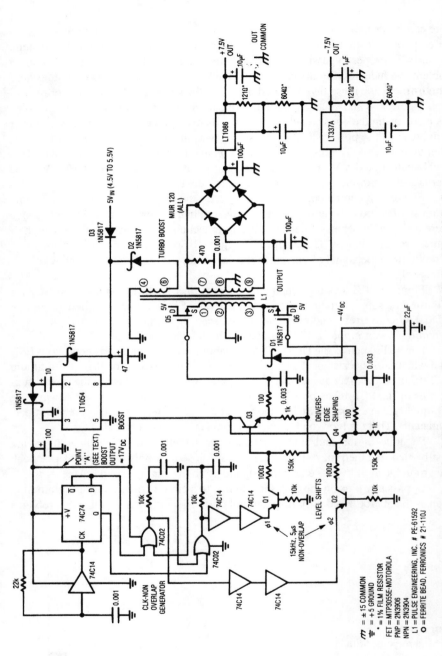

Figure 8-9. Low-noise converter for SCF use. (Linear Technology, Application Note 40, p. 6)

can arise. Pins connected to the summing junctions should also have bypass capacitors, if such pins are tied to analog ground in a single-supply system where noise is always present.

The power supplies used to power SCFs limit the maximum input and output signals to and from the filters. Thus, power supplies have a direct effect on the system dynamic range. For example, the LTC1064 type filter with ±7.5-V supplies provides ±5-V output swing. If the supplies are reduced to ±5 V, the output swing is ±3.3 V. For a filter with 450-mVp-p of output noise, these numbers translate to 87 dB and 83 dB of dynamic range, respectively.

8.4 Input Offset Problems

Typical offset voltages for the LTC1064 family through the four sections can be up to 40 mV or 50 mV. Although this might not be of concern for AC-coupled systems, it becomes important in DC-coupled applications. The anti-aliasing filter used before an A/D converter is a typical application where offset is an important concern.

For example, with a 5Vp-p input signal, the least significant bit (LSB) of an 8-bit A/D converter is approximately 20 mV, and one-half the LSB is about 10 mV. This implies that the use of a filter (or any device other than a straight wire) before an 8-bit A/D converter requires offset voltages below 10 mV. For a 12-bit converter, this provision requires 600 µV of offset at the A/D converter input.

Several methods of offset cancellation are common. The usual method with op amps is a pot that injects a correction voltage, as shown in Fig. 8–10. This method can correct the initial offset, but both the adjustment circuit and the CMOS op amps in the SCF have temperature coefficients that are not zero. If the circuit is used to correct offset at 25°C, the offset at other temperature will not be fully corrected. The major advantage of the Fig. 8–10 circuit is that the filter frequency response is not affected significantly. Figure 8–11 shows the time-domain response of the Fig. 8–10

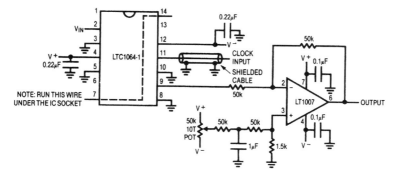

Figure 8–10. Elliptic filter with offset adjustment pot. (Linear Technology, Application Note 40, p. 7)

136 SIMPLIFIED DESIGN OF FILTER CIRCUITS

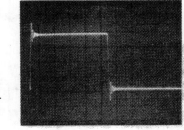

Figure 8–11.
Time-domain response of Elliptic filter. Linear Technology, Application Note 40, p. 7)

circuit. The rising and falling edge overshoot is typical of high-Q filters (both SCFs and active RC filters).

Figure 8–12 shows the circuit of an LTC1064-1 Elliptic filter with the same LT1007 output buffer amp used in Fig. 8–10. The Fig. 8–12 circuit also has an LT1012 op amp used as a servo to zero the filter offset. This arrangement can provide offsets of less than 100 µV, which are quite acceptable in a 12-bit system. The servo generates a low-frequency pole (at about 0.16 Hz) that can possibly interact with some signals of interest. Figure 8–13 shows the low-frequency square-wave response.

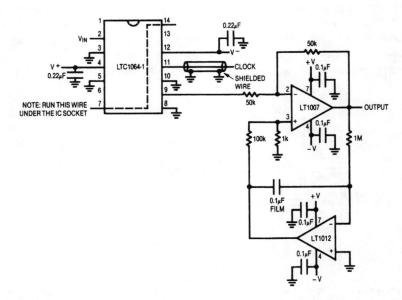

Figure 8–12. Elliptic filter with servo offset adjustment. (Linear Technology, Application Note 40, p. 8)

Practical Considerations for Switched-Capacitor Filters **137**

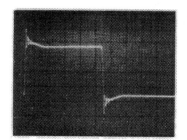

Figure 8–13.
Time-domain response of Elliptic filter with servo. (Linear Technology, Application Note 40, p. 8)

Figure 8–14 shows the distortion introduced for large-signal inputs to the Fig. 8–12 circuit at a frequency near the servo pole. The filter cutoff frequency is 100 Hz, and the input is a 1-Vrms 0.092-Hz sine wave.

Figure 8–15 shows the small-signal response to a sine-wave input at the servo system, but at lower amplitude and higher frequency (50 mVrms at 2 kHz, with a filter cutoff of 100 Hz). The small distortion introduced at this higher frequency is probably traceable to the high-frequency cutoff of the servo.

Figure 8–16 shows the servo response to the small-signal (50 mV) input at 0.092 Hz shown in Fig. 8–17. The filter cutoff remains at 100 Hz. The servo tracks the input but at a lower amplitude. As a result, the servo looks like a highpass filter to

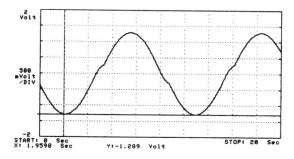

Figure 8–14.
Large-signal response of Elliptic filter with servo. (Linear Technology, Application Note 40, p. 8)

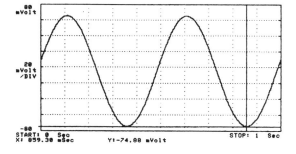

Figure 8–15.
Small-signal (50 mV, 2 Hz) response of Elliptic filter with servo. (Linear Technology, Application Note 40, p. 9)

138 SIMPLIFIED DESIGN OF FILTER CIRCUITS

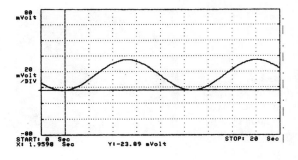

Figure 8-16.
Small-signal (50 mV, 0.092 Hz) response of Elliptic filter with servo. (Linear Technology, Application Note 40, p. 9)

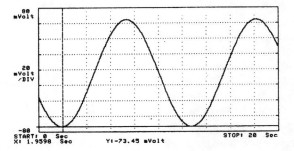

Figure 8-17.
Small-signal (50 mV, 0.092 Hz) input to Elliptic filter. (Linear Technology, Application Note 40, p. 9)

the input signal at input frequencies below the servo pole frequency. Servo offset nulling can be extremely useful in systems if the limitations described are tolerable.

8.5 Slew Limiting

The input stage of op amps in active RC filters or SCFs can be driven into slew limiting if the input-signal frequency is too high. Slew limiting is usually caused by a capacitive-load drive limitation in the internal circuits of the op amps. Most present-day SCFs are designed to avoid slew limiting (in most cases).

As an example, the LTC1064 filter has a typical slew rate of 10V/s. Because slew rate is a large-signal parameter, it also defines power bandwidth, which can be given as:

$$fp + SR/(6.28E_{op})$$

where fp is the full-power frequency (or power bandwidth), SR is slew rate, and Eop is the peak amplifier-output voltage.

For the LTC1064 operating at a VS of ±7.5 V, the device can swing ±5 V, or 10 V peak. Based on op-amp slew rate performance only, find the full-power frequency as follows:

$$fp = 10V/(6.28 \times 10^{-6} \times 10V) = 159 \text{ kHz}$$

Practical Considerations for Switched-Capacitor Filters **139**

This fp is sufficient for all but the most stringent SCF applications. (For a discussion of slew rate and bandwidth, read the author's *Simplified Design of IC Amplifiers,* Butterworth-Heinemann.)

8.6 Aliasing in SCFs

Because the SCF is based on a switching capacitor to generate variable filter parameters, it is (by definition) a sampled-data device. Like all other such devices,

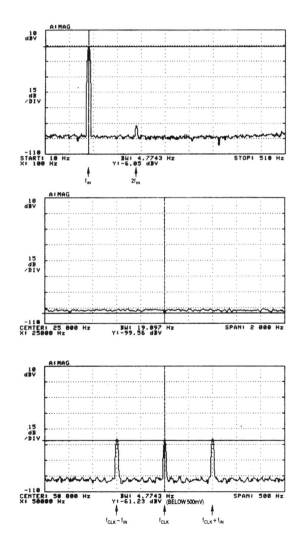

Figure 8-18.
Spectrum-analyzer displays of Elliptic filter with 100-Hz input. (Linear Technology, Application Note 40, p. 10)

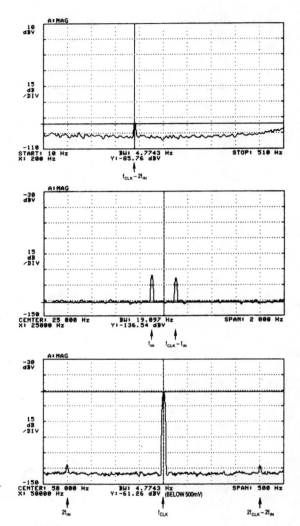

Figure 8-19.
Spectrum-analyzer displays of Elliptic filter with 24.9-kHz input. (Linear Technology, Application Note 40, p. 10)

the SCF is subject to aliasing. The simplest way to understand the effects of aliasing on SCFs is to study the spectrum analyzer displays shown in Figs. 8–18 through 8–21. In all four displays, fo = 500 Hz, fCLK = 50 kHz, VS = ±8 V. Also, all of these displays are for an Elliptic SCF using the LTC1064. Each illustration shows three spectrum segments as the SCF input signal varies from 100 Hz to 49.9 kHz.

Each figure shows a different input frequency and spectrum plots of the filter's passband (10 Hz – 510 Hz), the frequency spectrum around fCLK/2(25 kHz) and the

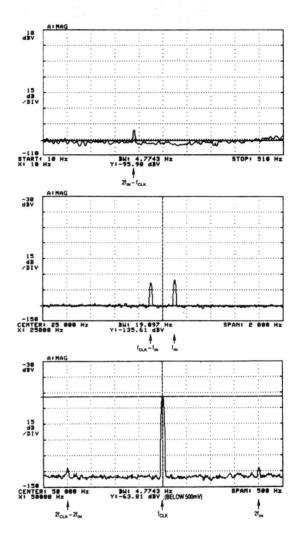

Figure 8–20.
Spectrum-analyzer displays of Elliptic filter with 25.1-kHz input. (Linear Technology, Application Note 40, p. 11)

frequency spectrum around fCLK(50 kHz). As shown, aliasing begins when the input signal passes the fCLK/2 threshold (25 kHz).

Figure 8–18 shows the LTC1064 in its normal mode of operation with the signal (100 Hz) within the passband of the filter. Note the small second-harmonic content (–80 dB) component that also appears in the passband. No signals are seen around fCLK/2. However, the original signal (100 Hz) appears attenuated at 49.9 kHz and 50.1 kHz. This is consistent with sampled-data theory and is important in

142 SIMPLIFIED DESIGN OF FILTER CIRCUITS

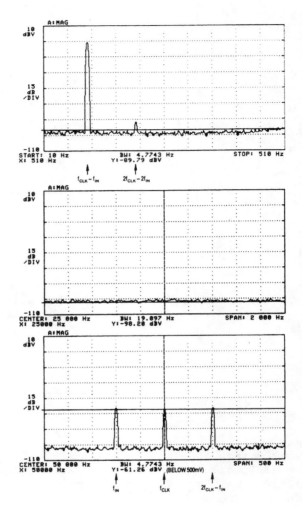

Figure 8–21.
Spectrum-analyzer displays of Elliptic filter with 49.4-kHz input. (Linear Technology, Application Note 40, p. 11)

some filter systems. For any signal input, there will be side lobes at fCLK ± fIN. These side lobes will be attenuated as shown.

Figure 8–19 shows the same series of spectrum plots for an input signal of 24.9 kHz. This signal is outside of the filter passband, so the signals are heavily attenuated. The signal seen at 200 Hz in this series of plots is actually the alias of the 49.8-kHz second harmonic of the input signal. The signals seen around 25 kHz are the 24.9-kHz input and the fCLK − fIN signals. The signals around and at 50 kHz are clock feedthrough, and the second-harmonic images of the 24.9-kHz input signal around the clock frequency.

Figure 8–20 shows spectral plots for an input signal of 25.1 kHz. The signals appear in the same locations as in Fig. 8–19, but for different reasons. The 200-Hz signal in Fig. 8–20 is the alias of the 50.2-kHz signal. The signals around 25 kHz are the input signal at 25.1 kHz and its alias at 24.9 kHz.

Figure 8–21 shows aliasing at the very worst. The input signal here is 24.9 kHz, which aliases back, in the filter passband, to 100 Hz and appears almost identical to the Fig. 8–18 100-Hz signal. The input signal appears at 49.9 kHz with its 50.1-kHz mirror image.

8.7 Choosing a Filter Response

The following is a summary of the factors to be considered when choosing a filter response for a particular application. This is based on use of the Linear Technology LTC1064 filters, but applies generally to all similar IC filters.

Figure 8–22 is a filter selection guide for the LTC1064 family. Figures 8–23 through 8–29 show both time-domain and frequency-domain responses for the various filter configurations. In all of the frequency-domain responses (Figs. 8–25 through 8–29), the cutoff frequency is 10 kHz, and VS is ±7.5 V. In the time-domain responses (Figs. 8–23 and 8–24) the cutoff frequency is 100 Hz, with a 10-Hz square-wave input. (The horizontal divisions are 20ms/div, with the vertical divisions 0.5 V/div.)

When comparing time-domain and frequency-domain responses, keep the following in mind. A time-domain response can be viewed on a scope as amplitude versus time. It is in the time domain that pulse overshoot, ringing, and distortion appear. The frequency domain (amplitude versus frequency) provides a look at the frequency response of a filter, and is monitored on a spectrum analyzer. From a practical, simplified-design standpoint, it is essential that the filter response be monitored both ways. This is because a filter with good frequency-domain response can have poor time-domain response and vice versa. For example, a filter with a square-wave input might provide the correct passband and stopband, but produce severe overshoot or ringing at the output. At the other extreme, a filter might pass a square wave without any distortion, but not at the desired cutoff frequency.

PART NUMBER	TYPE	PASSBAND RIPPLE	STOPBAND ATTENUATION	WIDEBAND NOISE 1Hz–1MHz	SNR	THD (1kHz) (NOTE 1)	SUPPLY VOLTAGE
LTC1064-1	Elliptic	±0.15dB	72dB @ 1.5f_c	150μV_{RMS}	1V_{RMS} Input = 76dB	−76dB	$V_S = \pm 5V$
				165μV_{RMS}	3V_{RMS} Input = 85dB	−70dB	$V_S = \pm 7.5V$
LTC1064-2	Butterworth	3dB	90dB @ 4f_c	80μV_{RMS}	1V_{RMS} Input = 82dB	−76dB	$V_S = \pm 5V$
				90μV_{RMS}	3V_{RMS} Input = 90dB	−70dB	$V_S = \pm 7.5V$
LTC1064-3	Bessel	3dB	60dB @ 5f_c	55μV_{RMS}	1V_{RMS} Input = 85dB	−76dB	$V_S = \pm 5V$
				60μV_{RMS}	3V_{RMS} Input = 94dB	−70dB	$V_S = \pm 7.5V$
LTC1064-4	Elliptic	±0.1dB	80dB @ 2f_c	120μV_{RMS}	1V_{RMS} Input = 78dB	−76dB	$V_S = \pm 5V$
				130μV_{RMS}	3V_{RMS} Input = 87dB	−70dB	$V_S = \pm 7.5V$

Note 1: These specifications from LTC data sheets represent typical values. Optimization may result in significantly better specifications. Call LTC for more details.

Figure 8–22. Filter selection guide for LTC1064 family. (Linear Technology, Application Note 40, p. 15)

144 SIMPLIFIED DESIGN OF FILTER CIRCUITS

Figure 8-23.
Time-domain response for LTC1064-1 filter. (Linear Technology, Application Note 40, p. 14)

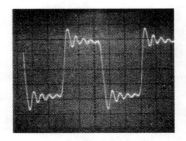

Figure 8-24.
Time-domain response for LTC1064-3 filter. (Linear Technology, Application Note 40, p. 14)

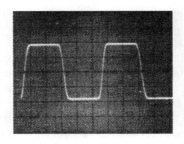

Figure 8-25.
Frequency-domain response for LTC1064-1 filter. (Linear Technology, Application Note 40, p. 14)

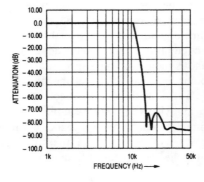

Figure 8-26.
Frequency-domain response for LTC1064-2 filter. (Linear Technology, Application Note 40, p. 14)

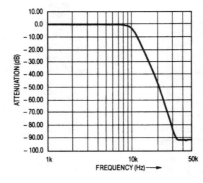

Figure 8-27.
Frequency-domain response for LTC1064-3 filter. (Linear Technology, Application Note 40, p. 14)

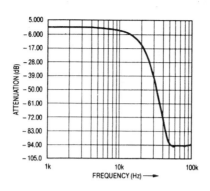

Figure 8-28.
Frequency-domain response for LTC1064-4 filter. (Linear Technology, Application Note 40, p. 14)

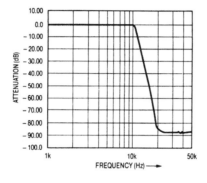

Figure 8-29.
Expanded comparison of LTC1064-1/LTC1064-4 rolloff from passband to stopband. (Linear Technology, Application Note 40, p. 15)

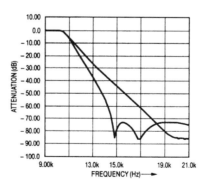

The frequency-domain responses shown in Figs. 8–25 through 8–28 are for the LTC1064 connected as 8th-order filters. Figure 8–29 shows an expansion of the comparison of the LTC1064-1 and LTC1064-4 rolloff from the passband to the stopband. The LTC1064-1 and LTC1064-4 are 8th-order Elliptic filters. The LTC1064-2 is Butterworth and the LTC1064-3 is Bessel. The rolloff from the passband to the stopband

146 SIMPLIFIED DESIGN OF FILTER CIRCUITS

is least steep for the LTC1064-3 Bessel. This is the price paid for the linear phase response that permits the Bessel to pass a square wave with good fidelity. Compare Fig. 8–23 (Elliptic) to Fig. 8–24 (Bessel). In a similar fashion, the LTC1064-2 Butterworth trades slightly worse transient response for steeper rolloff.

Figure 8–22 shows a wide variation in stopband-attenuation specifications for the various filters. Stopband attenuation is a measure of the filter's steepness of attenuation, and is a key specification for anti-aliasing filters found at the input of A/D converters. For example, the Elliptic response of the LTC1064-1 (with the corner frequency set at 10 kHz) would provide about 72 dB of attenuation to a 15-kHz input. The Butterworth response (LTC1064-2) provides about 27 dB under the same conditions, and the Bessel (LTC1064-3) produces only about 7 dB.

In general, use an Elliptic filter when the cutoff slope must be as steep as possible. A typical example is where the filter must discriminate among a series of continuous tones or frequencies, but where fidelity of signal is not critical. Use the Bessel where pulses are involved, and there must be no distortion (overshoot, ringing, etc.) or distortion must be at a minimum. Butterworth is a compromise between these extremes.

8.8 Noise in Filters (Active RC vs. SCF)

Figure 8–30 compares noise between an active RC-filter equivalent of the LTC1064-1 and the SCF itself. Both curves show typical peaking at the corner frequency. The active RC filter (trace B) is only slightly better than the SCF (trace A). This condition is true for most present-day SCFs in IC form. As the filter cutoff frequency increases, the SCF noise decreases, and the SCF becomes a better competitor to the active RC filter. Also, the noise in an SCF is constant, independent of bandwidth.

Figure 8–31 shows the frequency response of an 8th-order Bessel bandpass filter implemented with an LTC1064 as shown in Fig. 8–32. The response was made with a VIN of 1 V rms, an fCLK of 50 kHz, and a cutoff frequency of 1 kHz, using a buffered output. This filter has a Q of about nine, and a very linear phase response (in the passband) as shown in Fig. 8–33. The noise of the filter is shown in Fig. 8–34.

Figure 8–30.
Noise comparison between an active RC filter and an SCF. (Linear Technology, Application Note 40, p. 19)

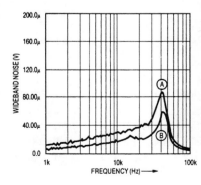

Figure 8-31.
Frequency response of an 8th-order Bessel bandpass filter. (Linear Technology, Application Note 40, p. 19)

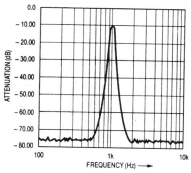

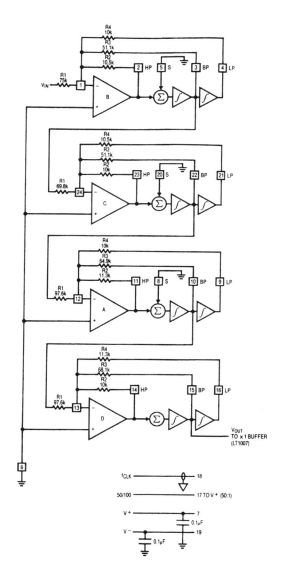

Figure 8-32.
8th-order Bessel bandpass filter using an LTC1064. (Linear Technology, Application Note 40, p. 20)

Figure 8-33.
Passband phase response of Bessel bandpass filter. (Linear Technology, Application Note 40, p. 20)

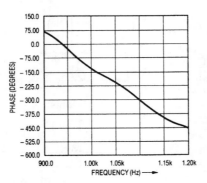

Figure 8-34.
Noise spectrum of Bessel bandpass filter. (Linear Technology, Application Note 40, p. 20)

Note that the noise bandshape is identical to the signal bandwidth of Fig. 8–31. This is not unusual because the bandpass filter is letting only the noise at a particular cutoff frequency through the filter. This is not clock feedthrough, and it is not peculiar to an SCF. In an active RC filter, or even an LC passive bandpass filter with these characteristics, noise appears "like a signal" at the center frequency of the bandpass filter.

8.9 Clock Problems in SCFs

A good stable clock is required to get good SCF performance. (This is generally the case with any clocked device!) The classic 555-timer type oscillators are not recommended for SCFs. Often, what appears as insufficient stopband attenuation, or excessive bandpass ripple, in SCFs is actually caused by poor clock sources.

Figure 8–35 shows the response of an LTC1064-1 set up to provide a cutoff frequency of 500 Hz. The clock was modulated (in the top curve measurement) to simulate about 50-percent clock jitter. The stopband attenuation at 750 Hz is about 42 dB

Practical Considerations for Switched-Capacitor Filters **149**

Figure 8-35.
Response of LTC1064-1 with and without clock jitter. (Linear Technology, Application Note 40, p. 21)

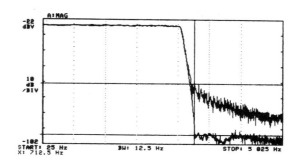

instead of the specified (LTC1064-1 data sheet) 68 dB. The bottom curve in Fig. 8-35 shows the response without clock jitter.

Similar graphs of the noise in Figs. 8-36 and 8-37 show the effect of clock jitter on the noise. (Figure 8-36 shows a 50-percent clock jitter, and Fig. 8-37 shows a clock jitter of less than 1 ns.) The wideband noise from 10 Hz to 1 kHz rises when a jittery clock is used from 156 Vrms to 173 μVrms. This is an increase of about 11 percent caused by poor clocking.

Figure 8-36.
Noise response of LTC1064-1 with 50 percent clock jitter. (Linear Technology, Application Note 40, p. 21)

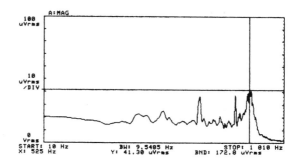

Figure 8-37.
Noise response of LTC1064-1 with low (1-ns) clock jitter. (Linear Technology, Application Note 40, p. 21)

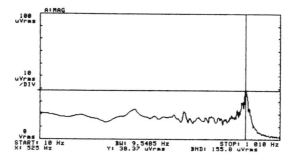

8.9.1 Synchronized Clocks

Whenever an SCF is used with another clocked device (such as an A/D converter) the clocks should by synchronized (preferably from the same source). In the case where the SCF is used ahead of the A/C to minimize aliasing, the A/D receives filtered data at a constant time, and ensures that the system has settled to the desired accuracy.

8.9.2 Clock Feedthrough

Clock feedthrough was one of the objections to early SCFs. Active RC filters were chosen in place of SCFs when noise (from any source) was a problem. This is still true today when noise is critical. In such cases, a simple (passive) RC filter can be added at the SCF output to reduce clock feedthrough. Such filters are known as post filters. See Fig. 8–3 for an example of a post filter (that could be added to the output of the LT1007CN).

Before going to an RC filter to eliminate clock feedthrough, keep in mind that the higher the clock-to-cutoff ratio, the easier it is to filter out feedthrough. Also, if the clock frequency is out of the band of interest, there is less of a need for an external RC filter.

Figure 8–18 (in the discussion on aliasing) shows the clock feedthrough at 50 kHz to be –61 dB. This is below 0 dB (which in this case is 500 mV). Clock feedthrough in Fig. 8–18 is about 400 µV. By inserting a passive RC filter (well outside the passband of the SCF) at the output of the SCF can reduces the feedthrough by a factor of 10.

Figure 8–38 shows a similar set of curves with a simple RC post filter on the output of the LT1007CN in Fig. 8–3. The RC values are 9.64 k and 3300 pF for a frequency of 5 kHz. Figure 8–38 shows that the clock feedthrough has been reduced to –82 dB below 500 mV (to 40 µV) when the post filter is used.

8.10 Bypass Capacitors for SCFs

Figure 8–39 shows the parasitic terms of a capacitor. When a capacitor is used to bypass a power supply line, the effect is to lower impedance at that point (typically the load). Because of the parasitic inductance and resistance in the supply lines (wires or PC traces), the impedance can be quite high. As frequency goes up, the inductive parasitic becomes a particular problem.

Even if there were no parasitics associated with a capacitor (the capacitor had capacitive reactance only), or if local regulation was used, bypassing would still be necessary. This is because no power supply or regulator has zero output impedance at high frequencies. The type of bandpass capacitor to be used is generally determined by the application (circuit frequency, cost, board space, and so on). Here are some useful generalizations about selecting bypass capacitors for use with SCFs and their power supplies.

Practical Considerations for Switched-Capacitor Filters **151**

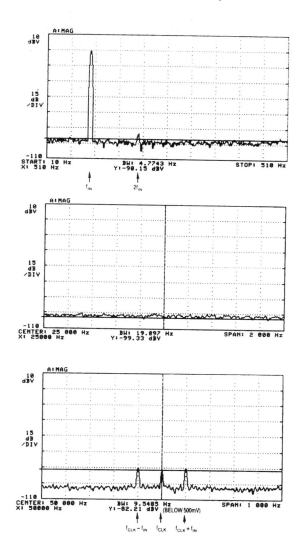

Figure 8–38.
Clock feedthrough with RC feedthrough-suppression filter. (Linear Technology, Application Note 40, p. 22)

In bypass applications, series resistance and series inductance (R and L in Fig. 8–39) are of primary importance when compared to dielectric absorption and leakage. Series R and L limit the capacitor's ability to damp transients and maintain low supply impedance. Also keep in mind that series L increases with frequency. Sometimes capacitors are rated as to equivalent series resistance or ESR. A low ESR is most desirable.

Bypass capacitors must often have a large capacitance value so that they can absorb long transients. Large-value capacitors usually mean electrolytic types (which

152 SIMPLIFIED DESIGN OF FILTER CIRCUITS

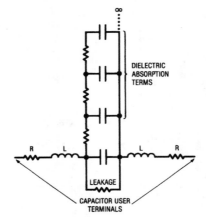

Figure 8-39.
Parasitic terms of a capacitor. (Linear Technology, Application Note 40, p. 26)

have large series R and L). Different types of electrolytics and electrolytic-non-polar combinations have decidedly different characteristics. To give you an example of these differences, Linear Technology has prepared a bypass capacitor test circuit shown in Fig. 8–40. Figures 8–41 through 8–45 show the results of testing various bypass capacitor combinations using the circuit of Fig. 8–40.

Figure 8–41 shows the results of an unbypassed power-supply line. Note that the B waveform sags and ripples badly.

Figure 8–42 shows the results when an aluminum 10-F electrolytic is used to bypass the line. This cuts quite a bit of the ripple, but still has problems.

Figure 8–43 shows the results when a tantalum 10-F capacitor is used for bypass. The response is considerably cleaner than with the aluminum bypass.

Figure 8–44 shows the results when a 10-µF aluminum electrolytic is connected in parallel with a 0.01-µF ceramic (non-polarized) capacitor. The results are even better than with the tantalum capacitor.

Combining electrolytics with non-polarized capacitors is a popular way to get good bypass response. However, there can be problems if the wrong combination is used. Figure 8–45 shows the results of combining the wrong paralleled dissimilar capacitors and supply-line parasitics. Note the excessive ringing response (which generally indicates undesired resonances). Always check the bypass responses in the experimental stages of an SCF (and most other ICs that require bypass capacitors!).

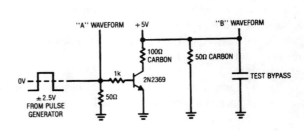

Figure 8-40.
Bypass capacitor test circuit. (Linear Technology, Application Note 40, p. 26)

Practical Considerations for Switched-Capacitor Filters **153**

Figure 8–41.
Response with unbypassed line. (Linear Technology, Application Note 40, p. 27)

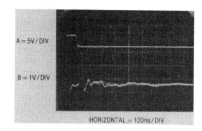

Figure 8–42.
Response with 10-µF aluminum capacitor. (Linear Technology, Application Note 40, p. 27)

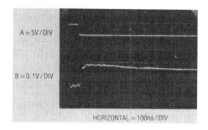

Figure 8–43.
Response with 10-µF tantalum capacitor. (Linear Technology, Application Note 40, p. 27)

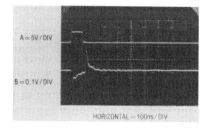

Figure 8–44.
Response with 10-µF aluminum capacitor in parallel with 0.01-µF ceramic. (Linear Technology, Application Note 40, p. 27)

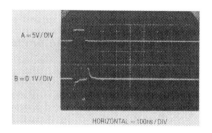

Figure 8–45.
Response with wrong paralleled dissimilar capacitors. (Linear Technology, Application Note 40, p. 27)

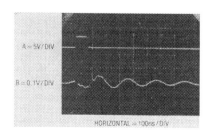

CHAPTER **9**

Active RC Filters Using Current-Feedback Amplifiers

This chapter is devoted to simplified design of active RC filters using current-feedback amplifiers or CFAs (sometimes called Norton amplifiers). CFAs are similar to op amps, but are closer to OTAs (operational transconductance amplifiers) because the CFA characteristics are controlled by an external current or voltage. However, the internal circuits and functions of CFAs are quite different from those of OTAs and op amps. For a thorough but dull discussion of CFAs, read the author's *Simplified Design of IC Amplifiers* (Butterworth-Heinemann).

9.1 Basic CFA Operation

With a CFA, there is no separate pin for control current. Any change in amplifier characteristics is set by current applied at the (+) noninverting and (–) inverting inputs. Some (but not all) manufacturers use a modified amplifier symbol such as that shown in Fig. 9–1 (which also shows a simplified version of the classic National Semiconductor LM3900). The current arrow between the inputs implies a current mode of operation. The symbol also signifies that current is removed from the (–) input and that the (+) input is a current input (which can control amplifier gain). The signal can be applied at either the (+) or (–) inputs.

9.2 Active RC Filter Basics

Before we get into actual circuit design, let us review active RC filter basics. Such filters have the obvious advantage of not requiring a clock, so there is no clock noise. Not so obvious is the fact that the performance characteristics of multistage RC filters are relatively insensitive to the tolerances of the RC components. This makes the performance of these filters easier to control in production runs. In many

156 SIMPLIFIED DESIGN OF FILTER CIRCUITS

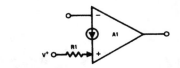

(a) Biasing From a "Noise-Free" Power Supply

Figure 9-1.
CFA bias considerations.
(National Semiconductor,
*Linear Applications
Handbook,* 1994, p. 181)

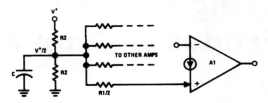

(b) Biasing From a "Noisy" Power Supply

cases where gain is needed in system design, it is relatively easy to also get frequency selectivity with active RC filters. However, the characteristics of the passive components must be considered carefully, as discussed next.

9.2.1 Resistor Characteristics

In active RC filter circuits, the passive components control performance. For this reason, carbon composition resistors are used mainly for room-temperature breadboarding or for final trimming of the more stable metal-film or wire-wound resistors. (Keep in mind that wire-wound resistors have inductance that changes with frequency. This characteristic can affect filter performance.)

9.2.2 Capacitor Characteristics

Capacitors present more of a problem in the range of values available, tolerance, and stability (with temperature, frequency, voltage, and time). For example, the disk ceramic type of capacitors are generally not recommended for active filter applications because of their relatively poor performance.

9.2.3 Impedance Scaling

Theoretically, it is possible to scale the impedance level of the passive components, to get an entirely different range of passbands and stopbands or to get more realistic values for components, without affecting filter characteristics. In a practical circuit using the CFAs in this chapter, excessive loading will result if the output resistor values are too small (typically less than 10 k). Such loading will reduce gain and could cause excessive output current (beyond the dissipation capabilities of the IC package).

Input resistor values present a different problem, primarily because of the DC bias current (typically about 30 nA) and because of the input impedance of amplifiers. The recommended solution to this problem is to reduce the impedance levels of all passive components to something less than 10 M. You get the best performance when all impedances are low, and when you do not try to get high gain HO, high Q (a Q greater than 50), and high center frequency (something greater than 1 kHz) simultaneously.

9.2.4 Sensitivity Functions

The term "sensitivity functions" is a measure of the effects of changes in the values of the passive components on filter performance. The term assumes infinite amplifier gain and relates the percentage change in the parameter of the filter, such as fo, Q, or HO, to a percentage change in a particular passive component. Small sensitivity functions (such as 1, ½, ¼) are most desirable.

Negative signs in sensitivity functions mean that an increase in the value of a passive component causes a decrease in that filter-performance characteristic. For example, if a bandpass filter has the following sensitivity function or factor applied:

$$S_{C3}^{\omega o} = -1/2$$

then, if C3 were to increase by 1 percent, ωo (the center frequency multiplied by 6.28 fc) would decrease by 0.5 percent.

9.3 Biasing CFA Active RC Filter Circuits

Figure 9–1 shows the basic biasing schemes for operation of CFA filter with a single supply. The connections for a noise-free and a noisy power supply are given. The basic connection is to use the (+) input to provide biasing. The power-supply voltage V+ is used as the DC reference to bias the output voltage of the CFA at about V+/2.

As shown in Fig. 9–1b, it might be necessary to remove AC components from a noisy power supply with a filter. This should remove noise from the filter output (at least the power-supply noise). One DC reference (with a filter as shown) can generally be used for all CFAs in the circuit. This is because there is essentially no signal feedback to the (+) input.

In all of the following CFA filter circuits, all of the CFAs are biased at V+/2. This allows the maximum AC voltage swing for any given DC power-supply voltage. The inputs to the following filters are also assumed to be at a DC level of V+/2 (for those circuits that are direct coupled).

9.4 Highpass Active RC Filter

Figure 9–2 shows a highpass RC filter using a single CFA. This circuit is biased using the (+) input of the CFA through R3. By making R3 equal to R2, and using a bias reference of V+/2, the output is set at V+/2. The input is capacitively coupled, so there are no further DC biasing problems.

158 SIMPLIFIED DESIGN OF FILTER CIRCUITS

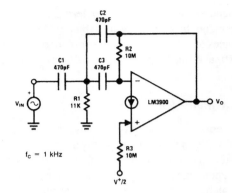

Figure 9-2.
Highpass RC filter using a CFA. (National Semiconductor, *Linear Applications Handbook,* 1994, p. 182)

The simplified design procedure for this filter is to select the passband gain HO, the Q, and the corner frequency fc. A Q value of 1 gives only a slight peaking (less than 2 dB) near the band edge. Peaking can be decreased with smaller Q values.

The slope of the skirt for this filter is 12 dB/octave (or 40 dB/decade). If the gain (HO) is unity, all capacitors will have the same value.

The given design factors are HO, Q, and ωc (which equals 6.28 fc). To find R1, R2, C1, C2, and C3, let C1 = C3 and choose a convenient starting value. Then:

$$R1 = \frac{1}{Q \omega c \, C1(2HO + 1)} \qquad (1)$$

$$R2 = \frac{Q}{c a o m e g a c \, C1}(2HO + 1) \qquad (2)$$

$$C2 = \frac{C1}{HO} \qquad (3)$$

Now let us design a highpass filter where:

$$HO = 1$$
$$Q = 10$$
$$fc = 1 \text{ kHz} (\omega c = 6.28 \times 10^3 \text{ rps})$$

Start by selecting C1 = 300 pF. Then, from equation (1)

$$R1 = \frac{1}{(10)(6.28 \times 10^3)(3 \times 10^{-10})(3)} = 17.7k$$

and from equation (2)

$$R2 = \frac{10(3)}{(6.28 \times 10^3)(3 \times 10^{-10})} = 15.9M$$

Active RC Filters Using Current-Feedback Amplifiers **159**

and from equation (3)

$$C2 = \frac{C1}{1} = C1$$

Now we must use *impedance scaling* to produce realistic values. Using the calculated value of 15.9 M for R2 makes R2 considerably larger than our desired 10 M (Section 9.2). We can reduce R2 to 10, and reduce R1 by the same scale factor (1.59), using

$$R1NEW = \frac{17.7 \times 10^3}{1.59} = 11.1k$$

and the capacitors are similarly reduced in impedance (increased in capacitance) as:

(C1 = C2 = C3)NEW = (1.59)(300)pF
C1NEW = 477pF

To complete the design, make R3 equal to R2 (10 M), resulting in a VREF of V+/2 to bias the output for larger signals.

In a practical circuit, the capacitor values should be adjusted to use standard values (470 pF instead of 477 pF).

9.5 Lowpass Active RC Filter

Figure 9–3 shows a lowpass RC filter using a single CFA. The resistor R4 is used to set the output bias level, and is selected after the other resistors have been established.

The simplified design procedure for this filter is to select HO, Q, and fc. Then find R1, R2, R3, R4, C1, and C2. Start by letting C1 be a convenient value. Then

$$C2 = KC1 \qquad (4)$$

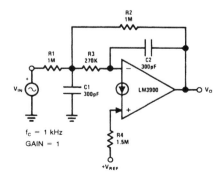

Figure 9–3.
Lowpass RC filter using a CFA. (National Semiconductor, *Linear Applications Handbook*, 1994, p. 182)

where K is a constant that can be used to adjust component values. For example, with K = 1, C1 = C2. Larger values of K can be used to reduce R2 and R3 at the expense of a larger value for C2. Then:

$$R1 = \frac{R2}{HO} \tag{5}$$

$$R2 = \frac{1}{2Q\omega cC1}\left[1 \pm \sqrt{1 + \frac{4Q^2(HO+1)}{K}}\right] \tag{6}$$

$$R3 = \frac{1}{\omega c^2 C1^2 R2(K)} \tag{7}$$

Now let us design a lowpass filter where:

$$HO = 1$$
$$Q = 1$$
$$fc = 1 \text{ kHz } (\omega c = 6.28 \times 10^3 \text{ rps})$$

Start by selecting C1 = 300 pF and K = 1, so C2 is also 300 pF.
 Now from equation (6)

$$R2 = \frac{1}{2(1)(6.28 \times 10^3)(3 \times 10^{-10})}\left[1\pm\sqrt{1+4(2)}\right]$$

$$R2 = 1.06 \text{ M}$$

Then from equation (5)

$$R1 = R2 = 1.06 M$$

and from equation (7)

$$R3 = \frac{1}{(6.28 \times 10^3)(3 \times 10^{-10})^2(1.06 \times 10^6)(1)}$$

$$R3 = 266k$$

To select R4, assume that the DC input level is 7 V and the DC output of the filter is also 7 V. This gives us the circuit of Fig. 9–4. Notice that an HO = 1 gives us not only equal resistor values (R1 and R2) but also simplifies the DC bias calculation as I1 = I2, and we have a DC amplifier with a gain of –1 (so if the DC input voltage increases 1 V, the output voltage decreases 1 V).

The resistors R1 and R2 (Fig. 9–3) are in parallel so the circuit can be simplified to that shown in Fig. 9–5, where the actual resistance values are given.

Active RC Filters Using Current-Feedback Amplifiers 161

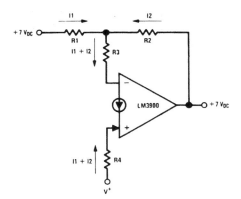

Figure 9-4.
Bias circuit for lowpass RC filter. (National Semiconductor, *Linear Applications Handbook,* 1994, p. 183)

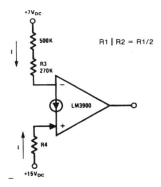

Figure 9-5.
Equivalent bias circuit for lowpass RC filter. (National Semiconductor, *Linear Applications Handbook,* 1994, p. 183)

The resistor R4 is given by

$$R4 = 2\left(\frac{R1}{2} + R3\right) + R3$$

or, using our values

$$R4 = 2\left(\frac{1M}{2} + 266k\right) \text{ about } 1.5M$$

9.6 Bandpass Active RC Filter

Figure 9-6 shows a bandpass RC filter using a single CFA. This circuit is suitable for low frequencies, low gain, and low Q (a Q of less than 10).

162 SIMPLIFIED DESIGN OF FILTER CIRCUITS

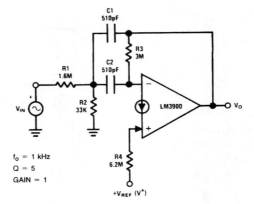

Figure 9–6.
Bandpass RC filter using a CFA. (National Semiconductor, *Linear Applications Handbook*, 1994, p. 183)

The simplified design procedure for this filter is to select HO, Q, and fo. Then find R1, R2, R3, R4, C1, and C2. Start by letting C1 = C2 and select a convenient value. Then:

$$R1 = \frac{Q}{HO\omega_o C1} \quad (8)$$

$$R2 = \frac{Q}{(2Q^2 - HO)\omega_o C1} \quad (9)$$

$$R3 = \frac{2Q}{\omega_o C1} \quad (10)$$

$$R4 = 2R3 \text{(for VREF = V +)} \quad (11)$$

Now let us design a bandpass filter where:

$$HO = 1$$
$$Q = 5$$
$$fo = 1kHz(\omega_o = 6.28 * 10^3 \text{ rps})$$

Start by selecting C1 = C2 = 510 pF
Then using equation (8)

$$R1 = \frac{5}{(6.28 \times 10^3)(5.1 \times 10^{-10})}$$

$$R1 = 1.57M$$

and using equation (9)

$$R2 = \frac{5}{[2(25)-1](6.28 \times 10^3)(5.1 \times 10^{-10})}$$

$$R2 = 3.13M$$

and using equation (10)

$$R3 = \frac{2(5)}{(6.28 \times 10^3)(5.1 \times 10^{-10})}$$

$$R3 = 3.13M$$

and using equation (11)

$$R4 = R3 \times 2$$
$$R4 = 6.2 \ M$$

9.7 Bandpass Active RC Filter with Two CFAs

Figure 9–7 shows a bandpass filter using two CFAs. This provides for a higher gain and higher Q (a Q between 10 and 50). The circuit uses two capacitors, and is similar to the single-CFA bandpass circuit of Fig. 9–6. The added CFA in Fig. 9–7 supplies a controlled amount of positive feedback to improve the response characteristics. The resistors R5 and R8 are used to bias the output voltage of the amplifiers at V+/2.

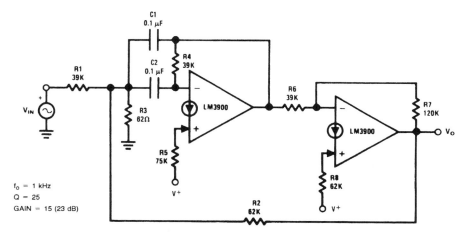

Figure 9–7. Bandpass RC filter using two CFAs. (National Semiconductor, *Linear Applications Handbook*, 1994, p. 184)

164 SIMPLIFIED DESIGN OF FILTER CIRCUITS

The simplified design procedure for this filter is to select Q and fo. Then find R1 through R8, C1, and C2. Note that R5 is chosen as twice the value of R4, and R8 must be selected after R6 and R7 have been assigned values. Start by letting C1 = C2 and choose a convenient starting value. Also choose a value for K to reduce the spread of element values or to optimize sensitivity. Typically, K should be greater than 1 but less than 10. Then:

$$R1 = R4 = R6 = \frac{Q}{\omega o C1} \tag{12}$$

$$R2 = R1 = \frac{KQ}{(2Q-1)} \tag{13}$$

$$R3 = \frac{R1}{Q^2 - 1 - 2/K + 1/KQ} \tag{14}$$

$$R7 = KR1 \tag{15}$$

$$HO = \sqrt{QK} \tag{16}$$

Now let us design a bandpass filter where:

$$Q = 25$$
$$fo = 1 \text{ kHz}$$

Start by selecting C1 = C2 = 0.1 μF and K = 3.
Then from equation (12)

$$R1 = R4 = R6 = \frac{25}{(6.28 \times 10^3)(10^{-7})}$$

$$R1 = R4 = R6 = 40 \text{ k}$$

from equation (13)

$$R2 = (40 \times 10^3) \frac{3(25)}{[2(25)-1]}$$

$$R2 = 61 \text{ k}$$

and from equation (14)

$$\{R3 = \frac{40 \times 10^3}{(25)^2 - 1 - 2/3 + \frac{1}{3(25)}}$$

$$R3 = 64 \text{ ohms}$$

from equation (15)

$$R7 = 3(40k) = 120k$$

and the gain is obtained from equation (16)

$$HO = \sqrt{25}(3) = 15(23dB)$$

To properly bias the first amplifier

$$R5 = 2R4 = 80k$$

and the second amplifier is biased by R8. Notice that the outputs of both amplifiers will be at V+/2. Therefore R6 and R7 can be paralleled and

$$R8 = 2(R6 \parallel 7)$$

or

$$R8 = 2\left[\frac{(40)(120) \times 10^3}{160}\right] = 59k$$

Notice that these values, to the closest standard resistor values, have been added to Fig. 9–7.

9.8 Bandpass Active RC Filter with Three CFAs (Bi-Quad)

Figure 9–8 shows a bandpass RC filter using three CFAs. This circuit provides a higher Q (a Q greater than 50) and reduces Q sensitivity to component variations even further. The circuit is a classic (originally used on analog computers) and is known as a Bi-Quad (because it can provide a transfer function that is "Quad"-ratic in both numerator and denominator, to give the "Bi").

The simplified design procedure for this filter is to select Q and fo. Then find R1 through R8, C1, and C2. Start by letting C1 = C2 and choose a convenient starting value. Also let 2R1 = R2 = R3 and choose a convenient starting value. Then:

$$R4 = R1(2Q - 1) \tag{17}$$

$$R5 = R7 = \frac{1}{\omega_o C1} \tag{18}$$

and for biasing the amplifiers we require

$$R6 = R8 = 2R5 \tag{19}$$

166 SIMPLIFIED DESIGN OF FILTER CIRCUITS

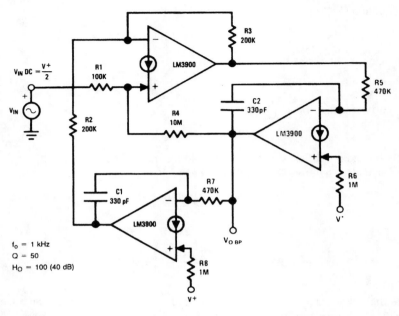

Figure 9-8. Bandpass RC filter using three CFAs. (National Semiconductor, *Linear Applications Handbook,* 1994, p. 185)

The mid-band gain is

$$HO = R4/R1 \qquad (20)$$

Now let us design a bandpass filter where:

$$Q = 50$$
$$fo = 1 \text{ kHz}$$

Start by selecting C1 = C2 = 330 pF and 2R1 = R2 = R3 = 360 k, and R1 = 180 k. Then from equation (17)

$$R4 = (1.8 \times 10^5)[2(50) - 1]$$
$$R4 = 17.7 \text{ M}$$

from equation (18)

$$R5 = R7 = \frac{1}{(6.28 \times 10^3)(3.3 \times 10^{-10})}$$

$$R5 = R7 = 483 \text{ k}$$

from equation (19)

$$R6 = R8 = 1M$$

From equation (20) the midband gain is 100 (40 dB). The value of 17.8 M for R4 is high, but can be lowered by scaling resistors R1 through R4 by a factor of 1.78 to give:

$$2R1 = R2 = R3 = \frac{360 \times 10^3}{1.78} = 200k, R1 = 100k$$

and

$$R4 = \frac{17.8 \times 10^6}{1.78} = 10M$$

These values, to the nearest 5 percent standard, are shown on Fig. 9–8.

CHAPTER **10**

Simplified Design Examples

This chapter is devoted to simplified design examples for a cross section of filters. All of the design information discussed thus far in this book applies to the examples in this chapter. However, each circuit has special design requirements that are discussed as necessary. The circuits in this chapter can be used the way they are, or by scaling component values (Section 9.2.3), as a basis for simplified design of similar filters.

Keep in mind that when a component value is altered to get a different filter characteristic, it is quite possible that the value for other components must be changed. This problem is especially critical in active RC filters, but also affects switched-capacitor filters. For example, the resistor-value relationships that determine filter Q usually affect gain and frequency. This is not a problem when the values are simply scaled (all changed by the same ratio) to get a different impedance (for input, output, etc.), but can produce drastic changes in filter characteristics when only one value is changed.

Finally, if you want to change the frequency of an SCF, change the clock frequency. (Or in some SCFs, you can change the clock-to-frequency ratio.) To change frequency of an active RC filter, you must change component values (which will probably change other characteristics).

10.1 DC-Accurate Filter for PLL Loops

Figure 10–1 shows an LTC1062 (Chapter 4) used as a loop filter for a PLL (phase-locked loop). Figure 10–2 shows the transient response of the PLL with a passive RC filter (displays A and B), and with the LTC1062 used as a loop filter (displays C and D). Figure 10–3 shows the passive lowpass RC filter used for displays A and B. Both the transient response and the jitter are drastically reduced when the LTC is used. In both cases, the output of the VCO is 6 kHz and the ÷N is 100.

Power supplies for the circuit of Fig. 10–1 are a single 5 V for the PLL and a ±5 V for the LTC1062. The PLL is a CMOS CD4046B. The LTC1062 can also be used with a single 5-V supply, but this results in some level shifting.

170 SIMPLIFIED DESIGN OF FILTER CIRCUITS

The PLL phase detector drives a diode-resistor combination to make the voltage at input R of the LTC1062 swing from one diode above ground to one diode be-

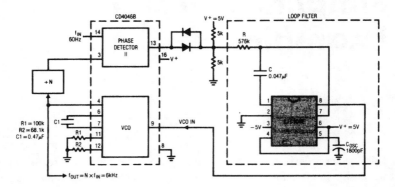

Figure 10-1. Loop filter for a PLL. (Linear Technology, Design Note 7, p. 2)

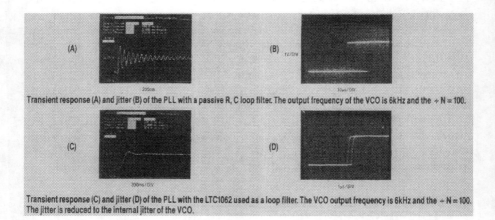

Figure 10-2. Transient response of PLL. (Linear Technology, Design Note 7, p. 2)

Figure 10-3.
Passive lowpass filter for PLL. (Linear Technology, Design Note 7, p. 2)

low the 5-V supply. The two 5-k resistors establish a maximum AC-impedance to keep the LTC1062 in the correct operating region, and to bias the VCO input at the midpoint when the phase detector switches to a three-state mode.

The circuit of Fig. 10–1 is designed for an input frequency (fIN) of 60 Hz, with a ±10 percent frequency range, and a divide-by-N of 100. The corner frequency (fc) of the LTC1062 is set at 9.5 Hz (which is about 1/6 of the fIN), and the internal clock is set for 2.4 kHz (about 252 times fc).

With the LTC1062 used as a loop filter, the settling time is 320 ms and the damping factor is about 0.7 when the input frequency is shifted (during test) from 54 Hz to 60 Hz. With the passive RC filter (Fig. 10–3), the settling time is 820 ms and the damping factor is about 0.1.

With the LTC1062 circuit, the VCO output jitter is about 1 s (with an fOUT of 6 kHz), measured over five periods. This jitter corresponds to a frequency error of 0.12 percent, and is quite adequate to drive the clock input of 0.3-percent-accurate SCFs (such as an LTC1059A or LTC1060A). With the passive RC filter, the jitter is 30 μs, which corresponds to a frequency error of about 18 percent.

To alter the circuit for other frequencies, make the fc of the LTC1062 greater than 1/6 of fIN, but less than 1/4 of fIN. Then set the internal (or external) clock frequency of the LTC at 150 to 250 times the desired fc.

10.2 Constant-Voltage Crossover Network

Figure 10–4 shows a constant-voltage crossover network with 12-dB/octave slopes. Using the values shown, calculations for the desired crossover frequency are reduced to a simple $1/(6.28 RC)$. Figure 10–5 shows the lowpass and highpass responses with a crossover frequency of 1 kHz. Note that the summed response (dashed lines) is flat.

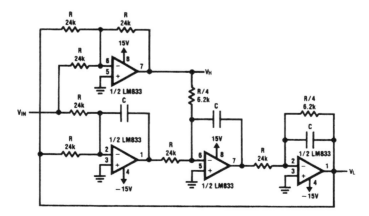

Figure 10–4. Constant-voltage crossover network. (National Semiconductor, *Linear Applications Handbook,* 1994, p. 826)

172 SIMPLIFIED DESIGN OF FILTER CIRCUITS

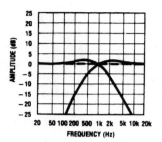

Figure 10–5. Lowpass and highpass responses of crossover network. (National Semiconductor, *Linear Applications Handbook*, 1994, p. 827)

10.3 Infrasonic and Ultrasonic Filters

Figure 10–6 shows an infrasonic filter (a 3rd-order Butterworth highpass) with a –3-dB at 15-Hz response. The attenuation at 5 Hz is over 28 dB, with 20-Hz information reduced by only 0.7 dB and 30-Hz information by less than 0.1 dB.

Figure 10–7 shows an ultrasonic filter (a 4th-order Bessel) which is down by 0.65 dB at 20 Hz and –3 dB at about 40 kHz. Rise time is limited to about 8.5 μs. This filter approximates a delay line within its passband, so complex in-band signals are passed through the circuit with negligible alteration of the phase relationships among the signals.

Figure 10–8 shows the amplitude response of the two filters (Figs. 10–6 and 10–7) connected in cascade. (When the two filters are cascaded, the lowpass should precede the highpass.)

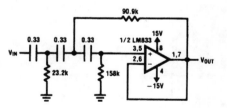

Figure 10–6. Infrasonic filter. (National Semiconductor, *Linear Applications Handbook*, 1994, p. 827)

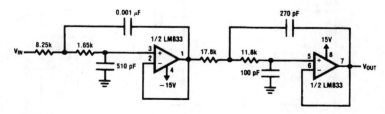

Figure 10–7. Ultrasonic filter. (National Semiconductor, *Linear Applications Handbook*, 1994, p. 827)

Simplified Design Examples **173**

Figure 10–8.
Amplitude response of infrasonic and ultrasonic filters. (National Semiconductor, *Linear Applications Handbook*, 1994, p. 828)

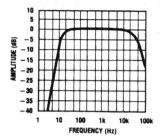

Both filters show a THD typically under 0.002 percent, and both must be driven from low-impedance sources (preferably under 100 ohms). Generally, 5-percent components will provide satisfactory results, but 1 percent values will keep the filter responses accurate and minimize mismatching between the two channels.

10.4 Lowpass Filters without DC Offset

Figure 10–9 shows a method of removing DC offset from a lowpass filter. The basic circuit is an 8th-order Butterworth lowpass filter with an fc of 10 kHz using a

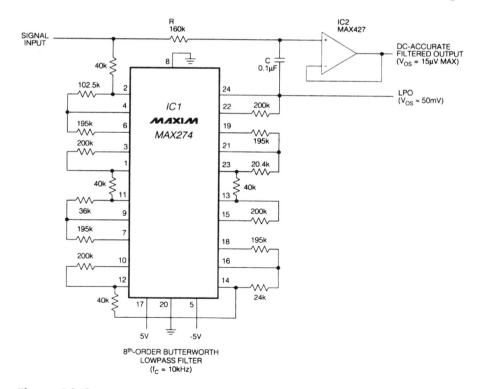

Figure 10–9. Removing DC offset from a lowpass filter. (*Maxim Engineering Journal*, Volume Ten, p. 10)

174 SIMPLIFIED DESIGN OF FILTER CIRCUITS

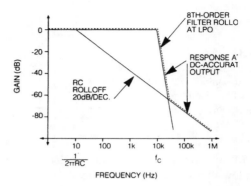

Figure 10-10.
Response of filter modified with RC network. (*Maxim Engineering Journal,* Volume Ten, p. 10)

Maxim MAX274. The circuit is modified by an RC network. Without the RC network, the output is offset by about 50 mV.

The RC values are selected so that the pole frequency is three decades below the filter cutoff frequency, or:

$$fc = 1/(6.28RC) = fc/1000$$

Using the values shown, the RC network frequency is:

$$1/(6.26 \times 160k \times 0.1\mu f) = 10Hz$$

Figure 10-10 shows that the RC network affects frequency response only in the stopband. Attenuation in the RC filter is sufficient so that the active-lowpass filter shape is maintained for frequencies between fc and the stopband. At higher frequencies, the RC network sets the filter rolloff rate at 20 dB/decade.

The RC network slows the process of nulling offsets to zero, but has no effect on circuit response to a step change in the DC input level. At DC and low frequencies, the DC-accurate output tracks the unfiltered input because R provides a signal path that bypasses the filter. At higher frequencies, C begins to conduct, allowing the DC-accurate output to track the filter lowpass output (LPO shown in Fig. 10-10). Assuming that the lowpass filter has unity gain and low ripple, the RC network has virtually no effect on filter gain or phase response. That is, the input signal and LPO swing together throughout most of the passband.

If desired, the RC output can be buffered with a low-offset op amp as shown in Fig. 10-9. The MAX427 guarantees an offset voltage of 5 µV typical and 15 µV maximum.

10.5 Highpass Filter with Synthetic Inductor

Figure 10-11 shows a method of simulating an inductor using two WTAs and capacitors. (If you are not familiar with WTAs, read the author's *Simplified Design of IC Amplifiers,* Butterworth-Heinemann.) From an operational standpoint, the circuit

Simplified Design Examples **175**

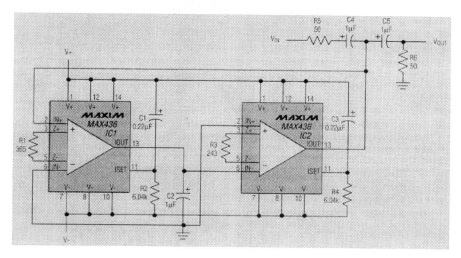

Figure 10-11. Simulating an inductor with two WTAs and capacitors. (*Maxim Engineering Journal,* Volume Thirteen, p. 16)

acts as a synthetic inductor, with one end connected to ground. This eliminates the need for an actual inductor (which can act as antennas to transmit and receive electromagnetic interference, or EMI). Actual inductors also have DC resistance that can affect filter characteristics.

Figure 10–12 shows the filter response. The filter is a 3rd-order Butterworth highpass with a maximum attenuation of 58 dB/decade. The slope decreases at the lower frequency because the synthetic-inductor Q is affected by the series resistance. (Comparable 1.25-mH inductors also have a resistance of about 53 ohms.) At 10 Hz, the attenuation for an ideal filter is –90 dB. For the circuit of Fig. 10–11, the attenuation is –80 dB.

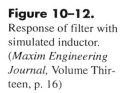

Figure 10-12. Response of filter with simulated inductor. (*Maxim Engineering Journal,* Volume Thirteen, p. 16)

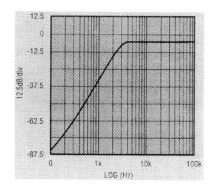

176 SIMPLIFIED DESIGN OF FILTER CIRCUITS

Capacitors C2 and C3 are bypass capacitors, but C2 is part of the simulated inductor. As a result, a change the value of C2 can affect filter characteristics (fc, passband, stopband, etc.).

The critical element in this circuit (from a simplified design standpoint) is setting the transconductance, or gm, of the WTAs. The gm for each WTA is set by external resistors (R1 and R3) according to the relationship gm = 8/R. The simulated inductance depends on the product of the two transconductances, so a change in the values of R1 and R2 can also affect filter characteristics.

If the characteristics are to be changed, leave the value of C2 at 1 µF, and change R1/R2 as necessary to get the desired characteristics. This can be done using computer simulation (Spice for example) or by actual measurement of the filter response. When the desired characteristics are obtained, the WTA transconductances must be checked for optimum value.

To find the optimum gm values, sweep the frequency at least one decade above and below the filter corner frequency and check each WTA output for peak voltage. Note that the two peaks might occur at different frequencies. At pin 13 of IC2 (this is the synthetic-inductor port) the peak value is set by the filter and cannot be changed. (A real inductor would produce the same peak.)

Adjust the peak of IC1 to match that of IC2. Let K equal the ratio of gm2 to gm1. Gain is proportional to gm, so divide gm1 by K and multiply gm2 by K. Finally, rerun the filter-characteristic measurements (or computer simulation) with the new gm values to verify that the peaks are equal, and the filter shape has not changed.

10.6 DC-Accurate Notch Filter

Figure 10–13a shows another method of simulating an inductor using two WTAs and capacitors. Figure 10–13b shows the equivalent passive filter. Figure 10–14 shows the filter response. The circuit of Fig. 10–13 is similar to that of Fig. 10–11, but the result is a 3.217-kHz 2nd-order notch.

Note that the DC path of the Fig. 10–13 circuit has no op amps, and therefore no DC offset. The path does not have a DC gain error, other than –6dB of attenuation caused by the RIN/RLOAD (R1/R2) divider. (This attenuation is absent for applications that omit the R2 termination.)

The AC path consists of a capacitor C1 and a synthetic inductor composed of two WTAs and their associated components. The result is an active circuit that emulates the passive filter of Fig. 10–13b.

As discussed in Section 10.5, simulating the inductance avoids the use of an actual inductor, which can (among other problems) act as a transmitting and receiving antenna for EMI. The equivalent inductance LEQ is determined by:

$$C/(gm1 \times gm2)$$

where gm1 and gm2 are the transconductances produced by IC1 and IC2. The inductance value can be large when gm1 times gm2 is substantially less than 1. However,

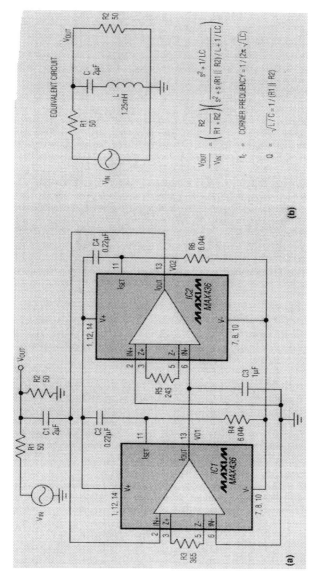

Figure 10–13. DC-accurate notch filter. (*Maxim Engineering Journal*, Volume Twenty, p. 13)

178 SIMPLIFIED DESIGN OF FILTER CIRCUITS

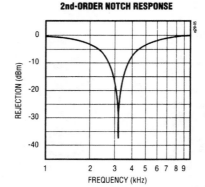

Figure 10–14.
Response of DC-accurate notch filter. (*Maxim Engineering Journal*, Volume Twenty, p. 14)

one end of the network must always connect to ground. Each gm is set by an external resistor (R3 or R5) according to the relationship gm = 8/R.

For optimum noise performance, the gm values should allow a full range of output swing for each WTA. Start with equal gm values and simulate the filter (in Spice) using "g" elements for the WTAs, or measure the actual filter characteristics.

Check the peak voltage amplitude at each WTA output while sweeping the frequency at least one decade above and below the filter corner frequency (3.217 kHz in this case). The peak value across the "inductor" (at pin 13 of IC2) is set by the filter and cannot be changed, so you adjust the peak value at pin 13 of IC1 to match. Let K equal the ratio of these peak values (V01pk/V02pk). Gain is proportional to gm, so divide gm1 by K and multiply gm2 by K. Then verify that the peaks are equal and that the filter shape has not changed.

Note that high-frequency error is dominated by parasitic capacitance between the output of the synthetic inductor and ground. Although small, this error increases when the parasitic reactance approaches the parallel combination of the source and load resistances. To minimize error in the frequency response, keep these resistances small with respect to the 3-k output impedance of the WTA.

10.7 Lowpass Filter for Anti-Aliasing

Figure 10–15 shows a Linear Technology LTC1064-1 connected as an 8th-order lowpass filter specifically designed for anti-aliasing. Figure 10–16 shows the response (which is Elliptic or Cauer, depending on the literature you read!). Figure 10–17 shows a comparison of the passband ripple between the LTC1064-1 filter and an RC active filter (the TL084). Figure 10–18 shows a comparison of other characteristics between the two filters.

Note that the circuit of Fig. 10–15 requires no external components, except for the two bypass capacitors to minimize power-supply noise and a 15-pF capacitor between pins 6 and 7. Also note that the clock-to-frequency ratio of 100:1 (fCLK = 4

Simplified Design Examples **179**

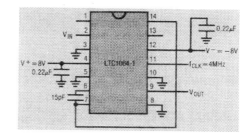

Figure 10-15.
Lowpass anti-aliasing filter. (Linear Technology, Design Note 16, p. 2)

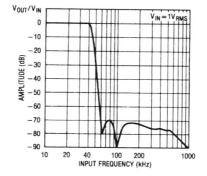

Figure 10-16.
Response of lowpass anti-aliasing filter. (Linear Technology, Design Note 16, p. 1)

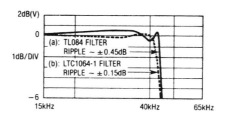

Figure 10-17.
Comparison of passband ripple. (Linear Technology, Design Note 16, p. 2)

	# EXT OP AMPS	# EXT R's, 1%	# EXT CAPS, 5%	TUNABLE	WIDEBAND NOISE, RMS[4]	DISTORTION $V_{IN}=1V_{RMS}, 3V_{RMS}$ (dB)	V_{OS} OUT (mV)[3]	I_{SUPPLY} (mA)	ATTENUATION AT 60kHz	MEASURED PASSBAND RIPPLE	TRIMM
RC Active TL084	16	31	8	No	111µV	−87, −87	55	33	65dB	±0.45dB	
LTC1064-1	None[1]	None	1	Yes	145µV	−70, −70	30	18	68dB	±0.15dB	

Note 1: An output inverting buffer (LT118) was used for driving cables during measurements.
Note 2: To obtain the ±0.45dB ripple for the TL084, 3 resistors were trimmed.
Note 3: The output offset voltage numbers are as measured by DVM with the input of the filter grounded.
Note 4: Measurement BW (2kHz–102kHz).

Figure 10-18. Comparison of anti-aliasing filter characteristics. (Linear Technology, Design Note 16, p. 2)

MHz, fc = 40 kHz). Both the LTC1064-1 circuit and the active RC filter can operate from ±7.5-V supplies, or from a single 15-V supply.

10.8 Communications Bandpass Filter with High Dynamic Range

Figure 10–19 shows two Linear Technology LTC1064s connected as a 16th-order bandpass filter specifically designed for communications applications (receiver IFs, Sonar, etc.). Figure 10–20 shows the response with an input of 2.2 Vrms. Note the flat response between 10 kHz and 100 kHz. Figure 10–21 shows the passband frequency response of the output buffer (LT1122).

The design combines an LTC1064-4 Elliptic lowpass filter and an LTC1064 quad highpass filter. The LTC1064-4 lowpass provides an attenuation greater than 70 dB at two times cutoff. The LTC1064 highpass is almost the mirror image of the lowpass.

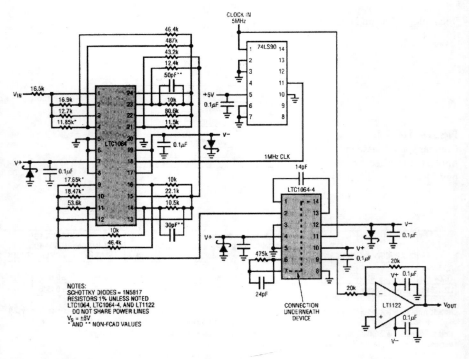

Figure 10–19. Communications bandpass filter. (Linear Technology, Design Note 37, p. 2)

Simplified Design Examples   **181**

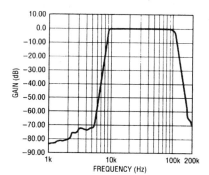

Figure 10–20.
Response of communications bandpass filter. (Linear Technology, Design Note 37, p. 1)

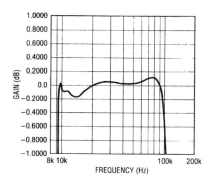

Figure 10–21.
Passband frequency response of output buffer. (Linear Technology, Design Note 37, p. 2)

The 74LS90 is used as a divide-by-5 to convert the 5-MHz external clock to 1 MHz for use by the LTC1064 highpass filter. The 5-MHz clock is applied directly to the LTC1064-1 lowpass. As a result, both filters have synchronous clocks (which is essential for good performance of any sampled-data device, including switched-capacitor filters).

Note that the 50-pF and 30-pF capacitors (shown with double asterisks) serve as RC filters on two of the LTC1064 highpass outputs. The capacitors roll off the response of the highpass filter well past the passband of the overall filter. This limits noise aliasing back to the filter passband. These capacitors might limit the overall tunability of the filter, or they might need a range-changing switch in some applications.

10.9 Monolithic 5-Pole Lowpass Filter

Figure 10–22 shows an LTC1063 clock-sweepable monolithic lowpass filter. Figure 10–23 shows how multiple LTC1063s can be synchronized. Figure 10–24

182 SIMPLIFIED DESIGN OF FILTER CIRCUITS

Figure 10–22.
Monolithic lowpass filter. (Linear Technology, Design Note 67, p. 1)

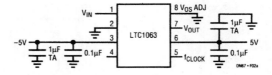

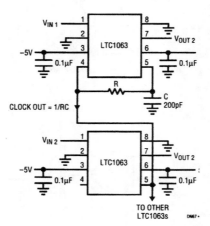

Figure 10–23.
Synchronizing multiple filters. (Linear Technology, Design Note 67, p. 1)

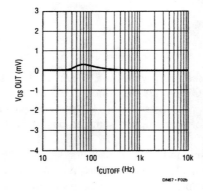

Figure 10–24.
DC offset versus cutoff frequency. (Linear Technology, Design Note 67, p. 1)

shows the output DC offset versus the cutoff frequency for the circuit of Fig. 10–22. This circuit approximates a 5-pole Butterworth lowpass filter.

The cutoff frequency is programmed by an internal or external clock. The clock-to-cutoff frequency ratio is 100:1. In the absence of an external clock, the internal oscillator can be used. An external resistor and capacitor (R and C in Fig. 10–23) set the internal clock frequency. The internal oscillator output is brought out at pin 4 so that it can be used as a synchronous master clock to drive other LTC1063s. Ten or more filters can be locked together to a single LTC1063 clock output as shown in Fig. 10–23.

The LTC1063 output DC-offset (typically 1 mV or less) is optimized for ±5-V supplies. Output offset is low enough to compete with discrete active RC filters using low-offset op amps. The circuit of Fig. 10–22 operates as a clock-sweepable lowpass filter, showing no more than 200 μV of total output-offset variation (Fig. 10–24) over three decades of cutoff frequency.

Figure 10–25 shows typical connections for measuring distortion-plus-noise and signal-to-THD-plus-noise ratio. Figure 10–26 shows a plot of distortion-plus-noise versus VIN, and illustrates that the filter can handle inputs to 4 Vrms (11.2 Vp-p) with less than 0.02 percent THD. With 4-V input, the dynamic range is only limited by distortion, and not by wideband noise. The signal-to-noise (S/N) ratio at 4-V input is 93 dB. Optimum S/N plus distortion (SINAD) is 83 dB. About 80 dB (0.01%) is achieved for input levels between 1 Vrms and 2.4 Vrms. Note that Fig. 10–26 shows the distortion-plus-noise versus input measured with a standard 1-kHz pure sine wave. The THD improves with increased power-supply voltage.

Figure 10–25.
Distortion, noise, and THD test circuit. (Linear Technology, Design Note 67, p. 2)

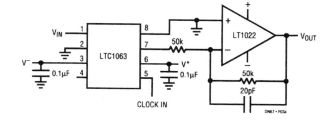

Figure 10–26.
Plot of distortion-plus-noise. (Linear Technology, Design Note 67, p. 2)

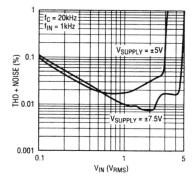

184 SIMPLIFIED DESIGN OF FILTER CIRCUITS

10.10 Notch Filter Using an Op Amp as a Gyrator

Figure 10–27 shows a 747 op amp connected to form a notch filter. The value of C1 sets the notch frequency in a range from 10 Hz to 10 kHz. For example, with a 0.01-μF value for C1, the notch frequency is 200 Hz. The remaining RC values remain the same throughout the range. However, it might be necessary to alter the value of R3 to get the desired response (Q, notch bandwidth, etc.).

10.11 Multiple Feedback Bandpass Filter

Figure 10–28 shows a 3403 op amp connected to form a bandpass filter using multiple feedback. A simplified design example is given for an fo of 1 kHz with a Q of 5. As shown, the R1, R2, and R3 values must be scaled when other characteristics are to be selected.

10.12 Filter with Both Notch and Bandpass Outputs

Figure 10–29 shows four sections of a 3403 op amp connected to form a biquad filter with both notch and bandpass outputs. The equations are simple, and there is a worked-out design example (an fo of 1 kHz with a bandwidth of 100 Hz). The gain for both the notch and bandpass outputs is unity (1). Notice that when R1, R2, and R3 are the same value and this value is a multiple of R (the frequency-determining resistance), the calculations are simplified.

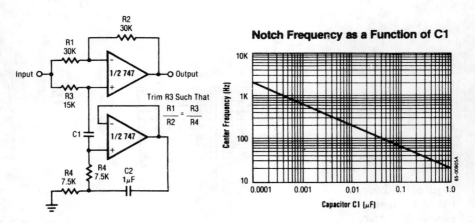

Figure 10–27. Notch filter with op-amp gyrator. (Raytheon Linear Integrated Circuits, p. 4-150)

Simplified Design Examples   **185**

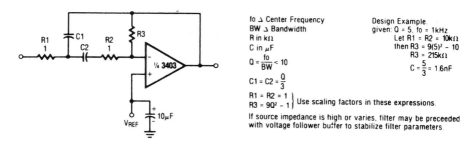

Figure 10-28. Multiple feedback bandpass circuit. (Raytheon Linear Integrated Circuits, p. 4-160)

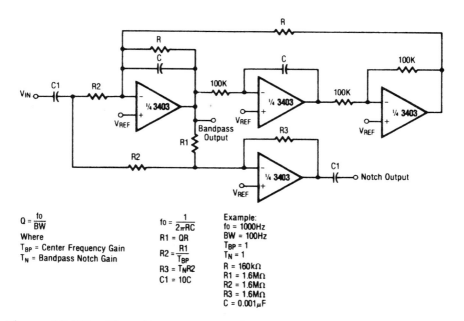

Figure 10-29. Filter with notch and bandpass outputs. (Raytheon Linear Integrated Circuits, p. 4-161)

10.13 Lowpass Butterworth Active RC Filter

Figure 10–30 shows four sections of a 4136 op amp connected to form a lowpass Butterworth filter with an fc of 400 Hz. Notice that all of the capacitors are 0.33 µF, and all must be changed or scaled by the same factor for other frequencies.

186 SIMPLIFIED DESIGN OF FILTER CIRCUITS

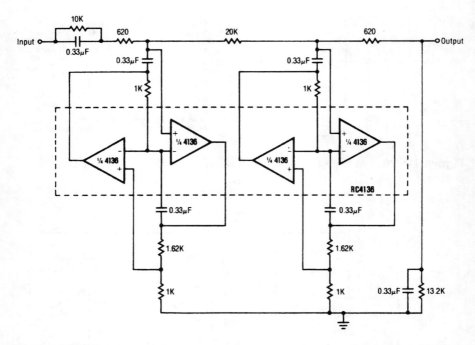

Figure 10–30. Lowpass Butterworth active RC filter. (Raytheon Linear Integrated Circuits, p. 4-172)

10.14 DC-Coupled Lowpass Active RC Filter

Figure 10–31 shows one section of a 4136 connected to form a lowpass filter with an fc of 1 kHz. Note that the frequency is set by the values of the capacitors and the 16-k resistors using the basic equation $fc = 1/(6.28RC)$. The frequency can be

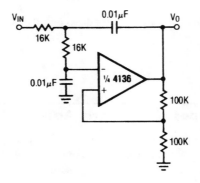

Figure 10–31.
DC-coupled lowpass active RC filter. (Raytheon Linear Integrated Circuits, p. 4-174)

changed by changing the capacitor values only. Both the input and output are direct coupled.

10.15 Universal State-Space Filter

Figure 10–32 shows four sections of an LM148 connected to form the classic universal state-space filter (with its equally classic, and complex, equations). These equations are best solved by computer (such as Spice simulation) and are thus not part of simplified design. However, no book on filters would be complete without a universal state-space circuit. If you decide to attempt this filter, follow the recommendations regarding fo and Q. (Use the bandpass output and adjust RO for the desired Q.)

10.16 Direct-Coupled Butterworth Filter

Figure 10–33 shows six sections of two LM148 op amps connected to form a Butterworth filter with direct-coupled inputs and outputs. The frequency of 1 kHz can be changed by changing the capacitor values (all of which are 0.001 µF for 1 kHz).

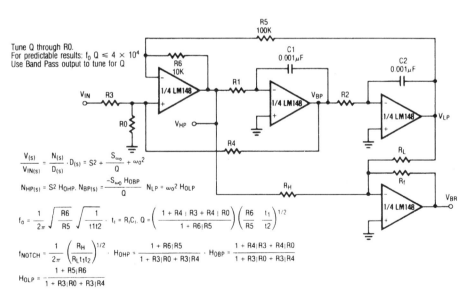

Figure 10–32. Universal state-space filter. (Raytheon Linear Integrated Circuits, p. 4-265)

188 SIMPLIFIED DESIGN OF FILTER CIRCUITS

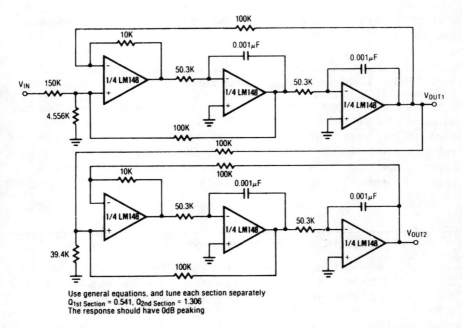

Figure 10–33. Direct-coupled Butterworth filter. (Raytheon Linear Integrated Circuits, p. 4-265)

10.17 Bi-Quad Notch Filter

Figure 10–34 shows three sections of an LM148 connected to form a bi-quad notch filter. Although these equations are complex, there is a design example (for a notch at 3 kHz with a Q of 5). These values can be altered and scaled for other frequencies and Q.

10.18 Elliptic Filter (Seven Section)

Figure 10–35 shows seven sections of two LM148s connected to form a classic active RC Elliptic filter. The filter response is also given, showing an fc of 1 kHz and an fs of 2 kHz. The calculations for this classic filter are best solved by computer. However, a design example is given. Notice that the calculations are somewhat simplified when all capacitors are made the same value (0.001 µF, in this case).

Simplified Design Examples   **189**

$$Q = \sqrt{\frac{R8}{R7}} \times \frac{R1C1}{\sqrt{R3C2R2C1}} \cdot f_0 = \frac{1}{2\pi}\sqrt{\frac{R8}{R7}} \times \frac{1}{\sqrt{R2R3C1C2}} \cdot f_{NOTCH} = \frac{1}{2\pi}\sqrt{\frac{R6}{R3R5R7C1C2}}$$

Necessary condition for notch: $\dfrac{1}{R6} = \dfrac{R1}{R4R7}$

Ex: f_{NOTCH} = 3kHz, Q = 5, R1 = 270K, R2 = R3 =20K, R4 = 27K, R5 = 20K, R6 = R8 = 10K, R7 = 100K,
C1 = C2 = 0.001μF

Better noise performance than the state-space approach

Figure 10–34. Bi-quad notch filter circuit. (Raytheon Linear Integrated Circuits, p. 4-266)

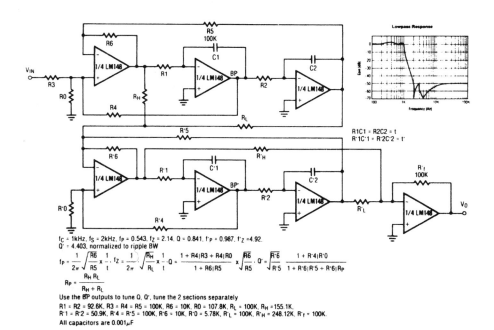

$R1C1 = R2C2 = t$
$R'1C'1 = R'2C'2 = t'$

f_C = 1kHz, f_S = 2kHz, f_P = 0.543, f_Z = 2.14, Q = 0.841, f'_P = 0.987, f'_Z = 4.92.
Q' = 4.403, normalized to ripple BW

$$f_P = \frac{1}{2\pi}\sqrt{\frac{R6}{R5}} \times \frac{1}{t}, \quad f_Z = \frac{1}{2\pi}\sqrt{\frac{R_H}{R_L}} \times \frac{1}{t} \cdot Q = \frac{1 + R4|R3 + R4|R0}{1 + R6|R5} \times \sqrt{\frac{R6}{R5}}, \quad Q' = \sqrt{\frac{R'6}{R'5}} \cdot \frac{1 + R'4|R'0}{1 + R'6|R'5 + R'6|R_P}$$

$R_P = \dfrac{R_H R_L}{R_H + R_L}$

Use the BP outputs to tune Q, Q', tune the 2 sections separately
R1 = R2 = 92.6K, R3 = R4 = R5 = 100K, R6 = 10K, R0 = 107.8K, R_L = 100K, R_H =155.1K.
R'1 = R'2 = 50.9K, R'4 = R'5 = 100K, R'6 = 10K, R'0 = 5.78K, R'_L = 100K, R'_H = 248.12K, R'_f = 100K.
All capacitors are 0.001μF.

Figure 10–35. Elliptic filter (seven section). (Raytheon Linear Integrated Circuits, p. 4-266)

190 SIMPLIFIED DESIGN OF FILTER CIRCUITS

10.19 Basic Piezo-Ceramic-Based Filter

Figure 10–36 shows the circuit for a highly selective bandpass filter that uses a resonant-ceramic element and a single amplifier. As shown by the response graph in Fig. 10–36, the ceramic element looks like a high impedance at all frequencies, except the resonant frequency (400 kHz in this case). At resonance, the ceramic element has a low impedance.

As shown in the schematic, the ceramic element has stray or parasitic capacitance that causes a slight rise in output at frequencies above 425 kHz. Typically, the A1 output is down about 20 dB at 300 kHz and 40 dB at 425 kHz. The parasitic capacitance can be minimized with a differential network, described next.

10.20 Piezo-Ceramic-Based Filter with Differential Network

Figure 10–37 shows a selective bandpass filter that is an improved version of the Fig. 10–36 circuit. With the Fig. 10–37 circuit, a portion of the input is fed to the noninverting input of A1. The RC network at this input looks like the ceramic-resonator impedance when the circuit is off null. As a result, A1 "sees" similar signals for out-of-band inputs. The high-frequency rolloff of Fig. 10–37 is smooth and about 20 dB deeper than the Fig. 10–36 filter at 475 kHz. The low-frequency side of resonance has similar characteristics at 375 kHz and below.

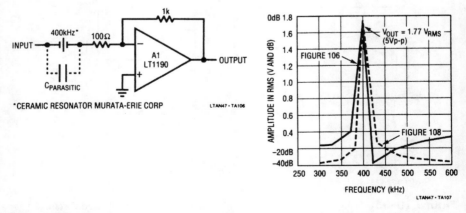

Figure 10–36. Basic Piezo-ceramic-based filter circuit. (Linear Technology, Application Note 47, p. 48)

Figure 10-37.
Piezo-ceramic-based filter with differential network. (Linear Technology, Application Note 47, p. 48)

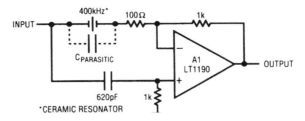

10.21 Basic Crystal Filter

The circuit shown in Fig. 10–38 replaces the ceramic element of Fig. 10–36 with a 3.57-MHz quartz crystal. The response of the Fig. 10–38 circuit shows almost 30-dB attenuation only a few kHz on either side of resonance.

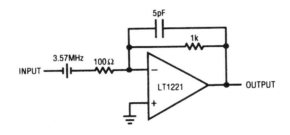

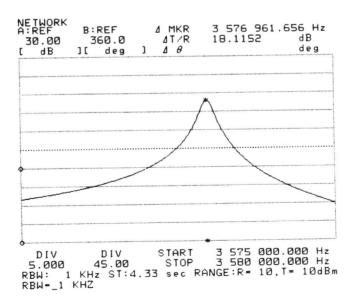

Figure 10-38. Basic crystal filter circuit. (Linear Technology, Application Note 47, pp. 48, 49)

10.22 Single 2nd-Order Filter Section

Figure 10–39 shows a single 2nd-order filter section using the MAX274/275 continuous-filter architecture. The circuit is a four-amplifier state-variable design. The on-chip capacitors and amplifiers, together with external resistors, form cascaded integrators with feedback to provide simultaneous lowpass and highpass filtered outputs.

The lowpass and bandpass frequencies, as well as filter Q, are determined by external resistor values, using the equations shown. No external capacitors are needed. On-chip capacitors are factory trimmed to provide better than 1-percent pole-frequency accuracy over the temperature range. One-percent-tolerance resistors provide 2-percent-accurate pole frequencies. Accurate filter Qs can also be obtained by compensating for amplifier-bandwidth limitations using the graphs that are provided on the data sheet.

10.23 Simple Highpass Butterworth Filter

Figure 10–40 shows a Harris Semiconductor HA-2544 video amplifier connected as an active RC filter with Butterworth response at 100 kHz. Notice that the response graph shows both frequency and phase-shift responses at frequencies from 10 Hz to 10 M. Frequencies other than 100 kHz can be selected using the basic frequency equation shown.

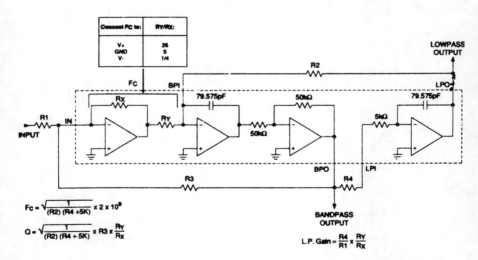

Figure 10–39. Single 2nd-order filter. (Maxim Applications and Products Highlights, p. 7–3)

Simplified Design Examples   **193**

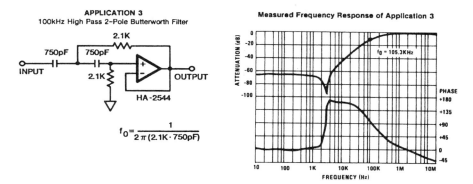

Figure 10–40.   Simple highpass Butterworth filter circuit. (Harris Semiconductor, Linear & Telecom ICs, 1991, p. 3-312)

10.24 Notch Filter with Adjustable Q

Figure 10–41 shows two LM110s connected to form a 60-Hz notch filter with adjustable Q. The Q value is set by adjustment of R4. Other frequencies can be selected using the equations.

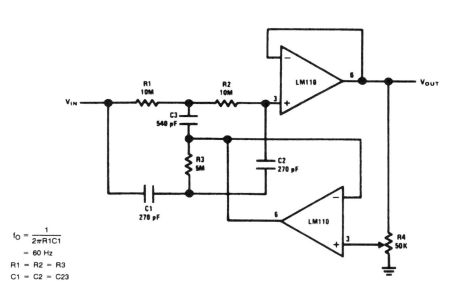

Figure 10–41.   Notch filter with adjustable Q. (National Semiconductor, Linear Applications Handbook, 1994, p. 84)

194 SIMPLIFIED DESIGN OF FILTER CIRCUITS

10.25 Easily-Tuned Notch Filter

Figure 10–42 shows an LM102 and an LM107 connected to form a tunable notch filter. The notch frequency is set by adjustment of C1. The frequency range is determined by the values of R4, C1, and C2.

10.26 Two-Stage Tuned Filter

Figure 10–43 shows two LM102s connected to form a two-stage tuned filter circuit. The frequency is set by the values of R1, R2, C1, and C2.

10.27 Basic Tuned-Filter Circuit

Figure 10–44 shows an LM101A connected to form a single-stage tuned filter. The output frequency is determined by the values of R1, R2, C1, and C2 as shown by the equations.

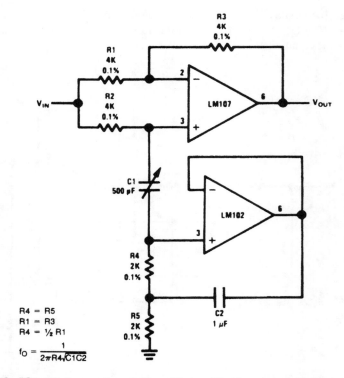

Figure 10–42. Easily tuned notch filter. (National Semiconductor, *Linear Applications Handbook,* 1994, p. 85)

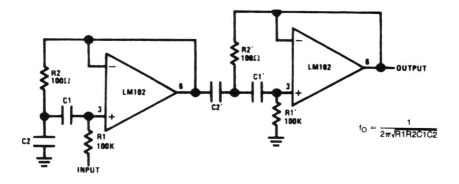

Figure 10–43. Two-stage tuned circuit. (National Semiconductor, *Linear Applications Handbook*, 1994, p. 85)

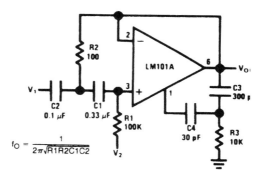

Figure 10–44.
Basic tuned-filter circuit.
(National Semiconductor,
*Linear Applications
Handbook*, 1994, p. 85)

10.28 Active RC Highpass Filter

Figure 10–45 shows an LM102 connected to form a basic highpass filter (input capacitors in series and input resistor in parallel). The values shown are for 100-Hz cutoff. Other frequencies can be selected with different values.

10.29 Active RC Lowpass Filter

Figure 10–46 shows an LM102 connected to form a basic lowpass filter (input capacitors in parallel and input resistor in series). The values are for a 10-kHz cutoff. Other frequencies can be selected with different values.

196 SIMPLIFIED DESIGN OF FILTER CIRCUITS

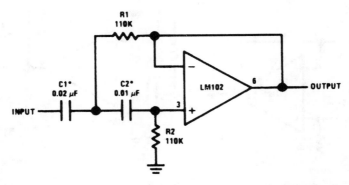

*Values are for 100 Hz cutoff. Use metalized polycarbonate capacitors for good temperature stability.

Figure 10–45. Active RC highpass filter. (National Semiconductor, *Linear Applications Handbook,* 1994, p. 87)

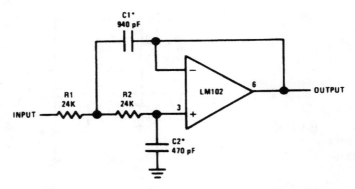

*Values are for 10 kHz cutoff. Use silvered mica capacitors for good temperature stability.

Figure 10–46. Active RC lowpass filter. (National Semiconductor, *Linear Applications Handbook,* 1994, p. 87)

10.30 Notch Filter with High Q

Figure 10–47 shows an LM110 connected to form a basic notch filter with high Q. The frequency range is determined by the values of R1 and C1 as shown by the equations.

10.31 Chebyshev Bandpass Filter

Figure 10–48 shows an LTC1064 switched-capacitor filter (Chapter 7) connected to form an 8th-order Chebyshev bandpass filter with a center frequency of 10.2 kHz and a bandwidth of 800 Hz.

Simplified Design Examples   **197**

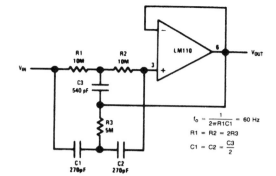

Figure 10–47.
Notch filter with high Q.
(National Semiconductor,
*Linear Applications
Handbook,* 1994, p. 1159)

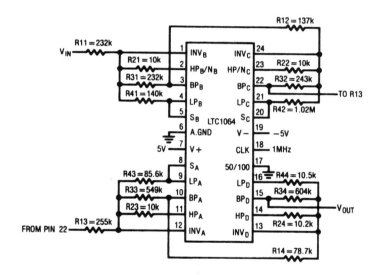

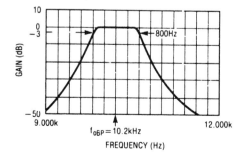

Figure 10–48.   Chebyshev bandpass filter. (Linear Technology, Application Note 27A, pp. 13, 15)

10.32 DC-Accurate Lowpass Bessel Filter

Figure 10–49 shows an LTC1050 and an LTC1062 (Chapter 4) connected to form a low-cost 7th-order 10-Hz lowpass filter, where amplitude and phase response closely approximate a true Bessel filter. The required clock frequency is 2 kHz, which yields a clock-to-cutoff frequency ratio of 200:1.

10.33 Simple Lowpass Filter

Figure 10–50 shows an LM107 connected to form a basic lowpass filter. As shown by the graph, the circuit has a 6-dB per octave rolloff, after a closed-loop 3-dB point that is defined by fc. Gain below the fc corner frequency is defined by the ratio of R3 to R1. The circuit can be considered as in integrator at frequencies well above fc. However, the time-domain response is that of a single RC, rather than an integral.

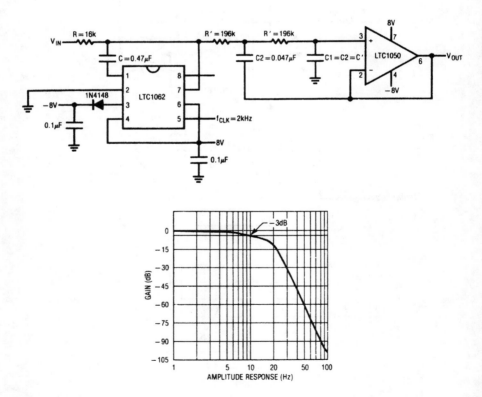

Figure 10–49. DC-accurate lowpass Bessel filter. (Linear Technology, Design Note 9, pp. 1, 2)

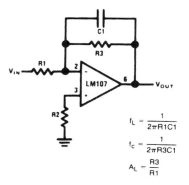 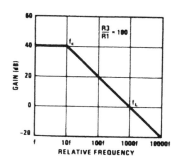

Figure 10–50. Simple lowpass filter. (National Semiconductor, *Linear Applications Handbook*, 1994, p. 22, 23)

R2 should be chosen so that its resistance is equal to the parallel combination of R1 and R3. This will minimize bias-current errors. The op amp should be compensated for unity-gain, or an internally compensated op amp should be used.

10.34 Wideband Highpass and Lowpass Filters

Figure 10–51 shows an LH0033 buffer connected to form a basic highpass filter with a 10-Hz cutoff frequency. A lowpass filter with the same frequency can be obtained by interchanging R1, R2, C1, and C2.

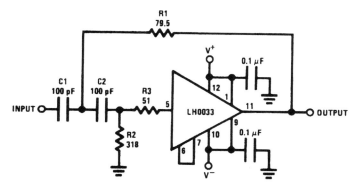

Figure 10–51. Wideband highpass and lowpass filters. (National Semiconductor, *Linear Applications Handbook*, 1994, p. 463)

10.35 Fed-Forward Lowpass Filter

Figure 10–52 shows two LM392 amplifier-comparators connected to form circuit that allows a signal to be rapidly acquired to final values, but also provides a long filtering constant. Such a characteristic is useful in multiplexed data-acquisition systems, and has been used in electronic infant-weighing scales where fast, stable readings of weight are needed (in spite of motion on the scale platform).

The point at which a filter switches from a short-time to a long-time constant is set by the 1-k pot. Normally, this pot is adjusted so that switching occurs at 90 to 98 percent of final value. However, the waveforms shown were obtained at a 70 percent trip point to show circuit operation.

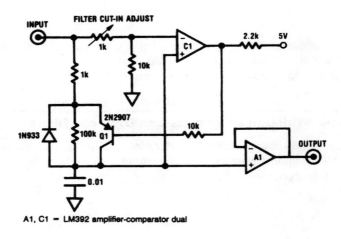

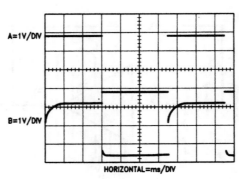

Figure 10–52. Fed-forward lowpass filter. (National Semiconductor, *Linear Applications Handbook,* 1994, p. 729)

10.36 4.5-MHz Notch Filter

Figure 10–53 shows an LH0033 buffer connected to form a notch filter. The frequency range is determined by the values of R1 and C1, as shown in the equations.

10.37 Spike Suppressor for Unregulated Power Supplies

Figure 10–54 shows a discrete-component circuit that suppresses transients in unregulated supplies. Zener D1 clamps the input voltage to the regulator, and L1 limits the current through D1 during the transient. The circuit will clamp 70-V 4-ms transients. The value of L1 = dvdt/I, where dv is the voltage by which the input transient exceeds the breakdown voltage of D1, dt is the duration of the transient, and I is the peak current of D1.

10.38 DC-Accurate Lowpass/Bandpass Filter

Figure 10–55 shows an LTC1050 and an LTC1062 (Chapter 4) connected to form a filter that extracts AC information from a DC plus AC signal. As shown by the response graph, this requires both lowpass and bandpass functions.

10.39 Simple Bandpass Filter

Figure 10–56 shows two sections of a 4136 op amp connected to form a bandpass filter with an fo of 1 kHz. An improved circuit is described next.

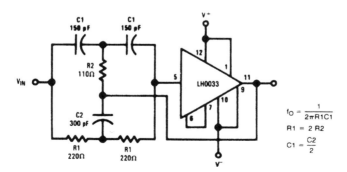

Figure 10–53. 4.5-MHz notch filter. (National Semiconductor, *Linear Applications Handbook*, 1994, p. 139)

Figure 10–54.
Spike suppressor for unregulated power supplies. (National Semiconductor, *Linear Applications Handbook*, 1994, p. 37)

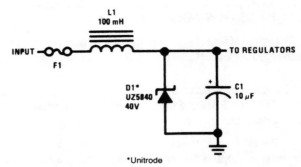

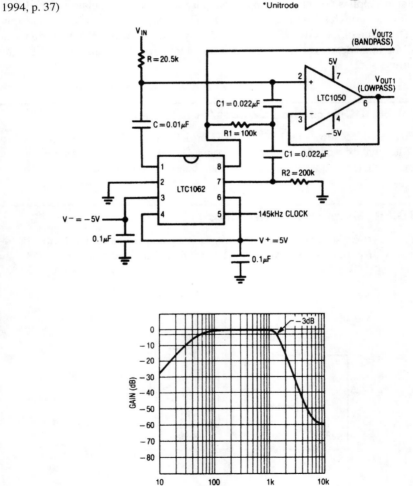

Figure 10–55. DC-accurate lowpass/bandpass filter. (Linear Technology, Design Note 9, p. 2)

Simplified Design Examples **203**

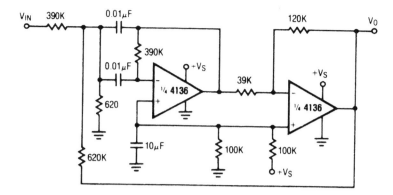

Figure 10-56. Simple bandpass filter. (Raytheon Linear Integrated Circuits, p. 4-175)

10.40 Bandpass Filter with High Q

Figure 10–57 shows two sections of a 3900 op amp connected to form a bandpass filter with an fo of 1 kHz, but with high Q (a Q of about 25).

10.41 Chebyshev Bandpass Filter with Single Clock

Figure 10–58 shows a MAX268 pin-programmable filter connected to provide a 10-kHz bandpass, centered at 50 kHz. The maximum passband ripple is 0.1 dB, with a gain of 1V/V (center frequency), and a Q of 4.2.

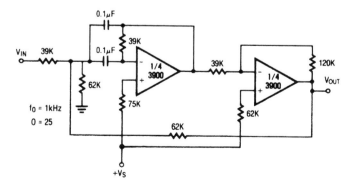

Figure 10-57. Bandpass filter with high Q. (Raytheon Linear Integrated Circuits, p. 4-274)

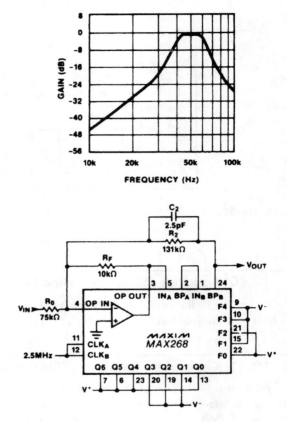

Figure 10-58.
Chebyshev bandpass
filter with single clock.
(Maxim New Releases
Data Book, 1992, p. 6-56)

10.42 Chebyshev Bandpass Filter with Two Clocks

The circuit of Fig. 10–59 is similar to that of Fig. 10–58 except that two clocks are required, but the only external component required is a 1.89-MHz crystal.

10.43 Dual-Tracking 3-kHz Lowpass Filter

Figure 10–60 shows two MAX263 pin-programmable filters connected to track each other and to provide a 3-kHz lowpass cutoff, with a Q of 1.3 and 0.5. The frequency response shown in the graph is that of a 4th-order Butterworth filter. The phase-shift response is also given.

Simplified Design Examples **205**

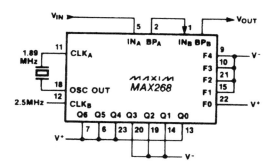

Figure 10–59.
Chebyshev bandpass filter with two clocks.
(Maxim New Releases Data Book, 1992, p. 6-56)

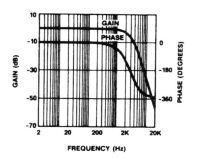

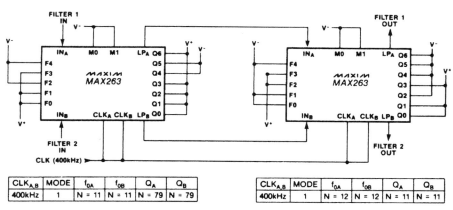

Figure 10–60. Dual-tracking 3-kHz lowpass filter. (Maxim New Releases Data Book, 1992, p. 6-88)

10.44 DC-Accurate Bessel Lowpass Filter

Figure 10–61 shows a MAX281 switched-capacitor filter connected to provide lowpass cutoff at 1 kHz. Other cutoff frequencies can be selected by altering the values of R and C as shown by the equations.

206 SIMPLIFIED DESIGN OF FILTER CIRCUITS

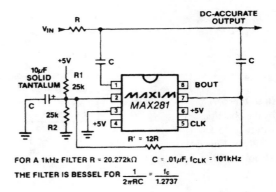

A. MAX281 WITH A SINGLE +5V SUPPLY

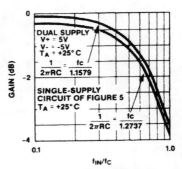

B. SINGLE- AND DUAL-SUPPLY PASSBAND FREQUENCY RESPONSE

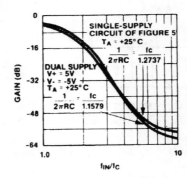

C. SINGLE- AND DUAL-SUPPLY STOPBAND FREQUENCY RESPONSE

Figure 10–61.
DC-accurate Bessel low-pass filter. (Maxim New Releases Data Book, 1992, p. 6-98)

10.45 Filtering AC Signals from High DC Signals

Figure 10–62 shows a MAX281 connected to remove undesired AC components from a DC voltage much higher than the operating voltage of the filter IC. This feature is possible because of the shunt architecture of the MAX281.

10.46 Switched-Capacitor Filters (MAX291-97)

Figure 10–63a shows MAX291-297 filters connected for single-supply operation. Figures 10–63b and 10–63c show the Elliptic filter response and pin descriptions, respectively. These filters are either 100:1 or 50:1 clock-to-corner-frequency ratio, as shown in Fig. 10–63d. (The MAX291-297 filters are similar to the filters described in Chapter 3).

The corner frequency depends directly on clock frequency, with a range from 0.1 Hz to 50 kHz. 100:1 is recommended for corner frequencies up to 25 kHz. Between 25 kHz and 50 kHz, the 50:1 ratio is recommended. For example, for a 20-kHz corner frequency, with a Bessel response, use the MAX292 and a clock of 2 MHz. The clock can be internal or external, with a maximum of 2.5 MHz.

For an internal clock, connect a capacitor between pin 1 and ground, using the equation:

$$\text{clock(kHz)} = \frac{10^5}{(3 \times \text{capacitor in pF})}$$

For example, for a clock of 20 kHz, use a 166-pF. This capacitance will produce a 200-Hz corner frequency.

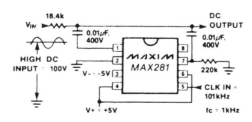
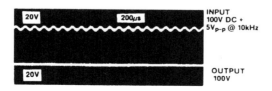

Figure 10–62.
Filtering AC signals from high DC signals. (Maxim New Releases Data Book, 1992, p. 6-98)

208 SIMPLIFIED DESIGN OF FILTER CIRCUITS

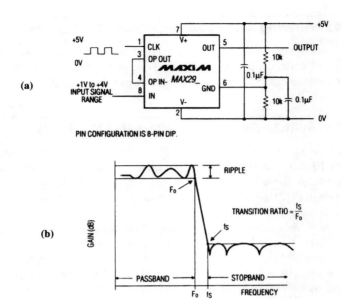

Figure 10–63. Switched-capacitor filters (MAX291-97). (Maxim New Releases Data Book, 1992, pp. 6-44, 6-45) (*Figure continued on next page.*)

Simplified Design Examples **209**

PART	RESPONSE SHAPE	CLOCK: CORNER-FREQUENCY RATIO
MAX291	Butterworth	100:1
MAX292	Bessel	100:1
MAX293	Elliptic (1.5 transition ratio)	100:1
MAX294	Elliptic (1.5 transition ratio)	100:1
MAX295	Butterworth	50:1
MAX296	Bessel	50:1
MAX297	Elliptic (1.5 transition ratio)	50:1

(d)

Figure 10–63. (Continued.)

10.47 Tabular Design of Butterworth Lowpass Filters

Figure 10–64 shows the basic circuit, and tabulated values, for using the uncommitted op amp in the MAX291-97 (Section 10.46) to form a lowpass filter. As shown by the table, the values given in the circuit of Fig. 10–64a will produce a corner frequency of 10 kHz. This circuit is intended for anti-aliasing applications, preceding the SCF in the MAX291-97, but can also be used as an active post-filter to reduce clock noise.

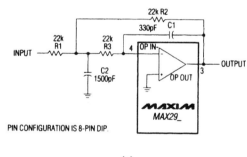

(a)

Corner Freq. (Hz)	R1 (kΩ)	R2 (kΩ)	R3 (kΩ)	C1 (F)	C2 (F)
100k	10	10	10	68p	330p
50k	20	20	20	68p	330p
25k	20	20	20	150p	680p
10k	22	22	22	330p	1.5n
1k	22	22	22	3.3p	15n
100	22	22	22	33n	150n
10	22	22	22	330n	1.5μ

NOTE: Some approximations have been made in selecting preferred component values.

Figure 10–64. Using an uncommitted op amp to form a lowpass filter. (Maxim New Releases Data Book, 1992, p. 6-45)

(b)

For Further Information

When applicable, the source for each circuit or table is included in the circuit or table title so that the reader may contact the original source for further information. To this end, the mailing address and telephone/fax number for each source is given in this section. When writing or calling, give complete information, including circuit title and description. Notice that all circuit diagrams and tables have been reproduced directly from the original source, without redrawing or resetting, by permission of the original publisher in each case.

AIE Magnetics
701 Murfreeboro Road
Nashville, TN 37210
(616) 244-9024

Analog Devices
One Technology Way
P.O. Box 9106
Norwood, MA 02062-9106
(617) 329-4700
Fax (617) 326-8703

Dallas Semiconductor
4401 S. Beltwood Parkway
Dallas, TX 75244-3292
(214) 450-0400

EXAR Corporation
48720 Kato Road
Freemont, CA 94538
(408) 434-6400
Fax (408) 943-8245

GEC Plessey Semiconductors
Cheney Manor
Swindon, Wiltshire
United Kingdom SN2 2QW
0793 518000
Fax 0793 518411

Harris Semiconductor
P.O. Box 883
Melbourne, FL 32002-0883
(407) 724-7000
(800) 442-7747
FAX (407) 724-3937

Linear Technology Corporation
1630 McCarthy Blvd.
Milpitas, CA 95035-7487
(408) 432-1900
Fax (408) 434-0507

Magnetics
Division of Spang and Company
900 East Butler
P.O. Box 391
Butler, PA 16003
(412) 282-8282

Maxim Integrated Products
120 San Gabriel Drive
Sunnyvale, CA 94086
(408) 737-7600
(800) 998-8800
Fax (408) 737-7194

Motorola, Inc.
Semiconductor Products Sector
Public Relations Department
5102 N. 56th Street
Phoenix, AZ 85018
(602) 952-3000

National Semiconductor Corporation
2900 Semiconductor Drive
P.O. Box 58090
Santa Clara, CA 95052-8090
(408) 721-5000
(800) 272-9959

Optical Electronics Inc.
P.O. Box 11140
Tucson, AZ 85737
(602) 889-8811

Philips Semiconductors
811 E. Arques Avenue
P.O. Box 3409
Sunnyvale, CA 94088-3409
(408) 991-2000

Raytheon Company
Semiconductor Division
350 Ellis Street
P.O. Box 7016
Mountain View, CA 94039-7016
(415) 968-9211
(800) 722-7074
Fax (415) 966-7742

Semtech Corporation
652 Mitchell Road
Newbury Park, CA 91320
(805) 498-2111

Siliconix Incorporated
2201 Laurelwood Road
Santa Clara, CA 95054
(408) 988-8000

Unitrode Corporation
8 Suburban Park Drive
Billerica, MA 01281
(508) 670-9086

Index

2nd-order filter section, single, 192
3 dB frequencies, 4–5
3-kHz lowpass filters, dual-tracking, 204–5
4.5-MHz notch filters, 201
5-Hz, filters (LTC1062) simple, 82–83
5-pole lowpass filters, monolithic, 181–83
7th-order Sallen/Key lowpass filters (MAX280), 78
15-V supplies (LTC1062), IC filters operated from, 84

AC signals, filtering from high DC signals, 207
Accuracy, 23
Active filters, 19–21
   multiple-feedback, 21
   Sallen-Key, 20–21
   universal state-variable, 21
Active lowpass filters, 47–63
   basic circuit functions for MAX270/271, 47–54
   digital threshold levels, 59
   filter performance, 59–60
   MAX270 control interface, 58
   MAX270/271, 47
   MAX271
      control interface, 58–59
      T/H (track-and-hold), 60
   power-supply configurations, 60–62
   programming
      cutoff frequency, 54–58
      without microprocessors, 62
   typical application (cascading), 63
Active RC filters, 155–67
   DC-coupled lowpass, 186–87
   bandpass, 161–63
   biasing CFA circuits, 157
   highpass, 157–59
   lowpass, 159–61
   lowpass Butterworth, 185–86
   with three CFAs, 165–67
   with two CFAs, 163–65
   versus SCFs, 129–30
Active RC highpass filter, 195–96
Active RC lowpass filter, 195–96
Active RC vs. SCF, 146–48
Active-filter, characteristics, 21
AGC (automatic-gain-control) circuits, 43
Aliasing, 25
   in SCFs, 139–43
Allpass filters, 11
Amplifiers, Norton, 155–67
Amplitude response defined, 2
Anti-aliasing, 74
   lowpass filter for, 178–80
Attenuation
   filter, 73
   rate near cutoff, 11

214 Index

Band-reject filters, 7–8
Bandpass active RC filters, 161–63
  with three CFAs (Bi-Quad), 165–67
  with two CFAs, 163–65
Bandpass filters, 2, 5–7, 113–28
  DC-accurate lowpass, 201
  cascading bandpass filters in Mode 1, 126–27
  cascading bandpass filters in Mode 2, 127–28
  cascading identical IC filters, 125–26
  cascading more than two identical 2nd-order sections, 128
  cascading non-identical IC filters, 122–25
  Chebyshev, 196–97
  gain and phase relationships of IC filters, 119–21
    bandpass gain versus Q, 119–20
    constant Q versus constant bandwidth, 121
    phase shift in filters, 121
  with high dynamic ranges, 180–81
  with high Qs, 203
  in Mode 1, 126–27
  in Mode 2, 127–28
  multiple feedback, 184
  selective clock-sweepable, 84
  simple, 201–3
  with single clocks, 203–4
  tabular design of, 113–28
  with two clocks, 204–5
  using tables, 113–19
    Butterworth tables, 113–19
    Chebyshev tables, 113–19
Bandpass gain versus Q, 119–20
Bandpass outputs, filter with, 184–85
Bandwidth, constant Q versus constant, 121
Bessel filters, 16–18
  DC-accurate lowpass, 198

Bessel lowpass filter, DC-accurate, 205–6
Bi-Quad
  bandpass active RC filter with three CFAs, 165–67
  notch filter, 188
Biasing CFA active RC filter circuits, 157
Buffered cascade lowpass filter (LTC1062), 79–80
Buffers, internal, 73
Butterworth
  direct-coupled filters, 187–88
  filters, 13
  highpass filter, 175
  lowpass active RC filter, 185–86
  lowpass filters, 182, 209
  tables, 113–19
Bypass capacitors for SCFs, 150–53

Capacitors
  characteristics, 156
  programmable, 51–53
  for SCFs, 150–53
  values, 71–72
Cascading, 3, 63
  bandpass filters in Mode 1, 126–27
  bandpass filters in Mode 2, 127–28
  identical IC filters, 125–26
  more than two identical 2nd-order sections, 128
  non-identical IC filters, 122–25
Center frequencies, 5, 30–31
CFAs (current-feedback amplifiers)
  active RC filter circuits biasing, 157
  active RC filters using, 155–67
  bandpass active RC filter with three, 165–67
  bandpass active RC filter with two, 163–65
Chebyshev
  bandpass filters, 196–97

Index **215**

with single clock, 203–4
with two clocks, 204–5
filters, 13–16
tables, 113–19
Circuits
automatic-gain-control (AGC), 43
board areas, 25
functions for MAX280/LTC 1062, 65–68
Clock
Chebyshev bandpass filter with single, 203–4
divider ratios, 68
problems in SCFs, 148–50
Clock-sweepable
bandpass filter (LTC1062), 84
pseudo-bandpass notch filter (LTC1062), 84
Clock-tunable notch filter (LTC1062), 84
Clocks
Chebyshev bandpass filter with two, 204–5
external, 30, 71
Communications bandpass filter with high dynamic range, 180–81
Component count, 25
Configurations, basic filter, 31
Constant-voltage crossover networks, 171–72
Continuous lowpass filters, 47–63
basic circuit functions for MAX270/271, 47–54
digital threshold levels, 59
filter performance, 59–60
MAX270 control interface, 58
MAX270/271, 47
MAX271
control interface, 58–59
T/H (track-and-hold), 60
power-supply configurations, 60–62
programming
cutoff frequency, 54–58

without microprocessors, 62
typical application (cascading), 63
Control interface
MAX270, 58
MAX271, 58–59
Corner frequency, 41
Cost, 24
Crossover networks
constant-voltage, 171–72
defined, 11
Crystal filter, 191
Current, quiescent, 53–54
Cutoff
attenuation rate near, 11
characteristic, 11
Cutoff frequencies, 5, 41
programming, 54–58
accuracy, 58

DC offset, lowpass filters without, 173–74
DC signals, filtering AC signals from high, 207
DC-accurate
Bessel lowpass filter, 205–6
cascaded lowpass filter (LTC1062), 78–79
filter for PLL loops, 169–71
lowpass Bessel filter, 198
lowpass/bandpass filter, 201
notch filter, 176–78
DC-coupled lowpass active RC filter, 186–87
Design effort, 25
Design examples, simplified, 169–209
4.5-MHz notch filters, 201
DC-accurate
Bessel lowpass filter, 205–6
filter for PLL loops, 169–71
lowpass Bessel filter, 198
lowpass/bandpass filter, 201
notch filter, 176–78

216 Index

Design examples, simplified (*Cont,*):
  DC-coupled lowpass active RC
      filter, 186–87
  active RC
      highpass filter, 195–96
      lowpass filter, 195–96
  bandpass filter with high Q, 203
  basic
      crystal filter, 191
      piezo-ceramic-based filter, 190
      tuned-filter circuit, 194–95
  bi-quad notch filter, 188
  Chebyshev bandpass filters, 196–97
      with single clocks, 203–4
      with two clocks, 204–5
  communications bandpass filter
      with high dynamic range,
      180–81
  constant-voltage crossover
      networks, 171–72
  direct-coupled Butterworth filter,
      187–88
  dual-tracking 3-kHz lowpass filter,
      204–5
  easily-tuned notch filter, 194
  elliptic filters, 188–89
  fed-forward lowpass filters, 200
  filter with notch and bandpass
      outputs, 184–85
  filtering AC signals from high DC
      signals, 207
  highpass filter with synthetic
      inductor, 174–76
  infrasonic filters, 172–73
  lowpass Butterworth active RC
      filters, 185–86
  lowpass filters
      for anti-aliasing, 178–80
      without DC offset, 173–74
  monolithic 5-pole lowpass filter,
      181–83
  multiple feedback bandpass filter,
      184

  notch filters
      with adjustable Qs, 193
      with high Qs, 196–97
      using op amps as gyrators, 184
  piezo-ceramic-based filter with
      differential network, 190–91
  simple
      bandpass filters, 201–3
      highpass Butterworth filters,
      192–93
      lowpass filter, 198–99
  single 2nd-order filter section, 192
  spike suppressor for unregulated
      power supplies, 201
  switched-capacitor filters
      (MAX291-97), 207–9
  tabular design of Butterworth
      lowpass filters, 209
  two-stage tuned filter, 194
  ultrasonic filters, 172–73
  universal state-space filter, 187
  wideband
      highpass filters, 199
      lowpass filters, 199
Design hints for all modes of
    operation, 31–32
Differential network,
    piezo-ceramic-based filter with,
    190–91
Digital threshold levels, 59
Direct-coupled Butterworth filter,
    187–88
Divider ratios, clock, 68
Dual-tracking 3-kHz lowpass filter,
    204–5
Dynamic range, communications
    bandpass filter with high, 180–81

Electronic filters, 1–25
  3 dB frequencies, 4–5
  filter order, 3–4
  filter Q, 4–5

functions, 1
introduction to, 1–25
  active filters, 19–21
  best filter in world, 23–25
  classic filter functions, 13–18
  passive filters, 18–19
  switched-capacitor filters, 21–23
log scales, 4–5
phase shift, 1–3
poles, 3–4
response, 1–3
transfer functions, 1–3
types, 5–11
  allpass filters, 11
  band-reject filters, 7–8
  bandpass filters, 5–7
  filter approximations, 11–13
  high-cut filters, 9
  highpass filters, 9–11
  lowpass filters, 8–9
  notch filters, 7–8
  phase-shift filters, 11
zeroes, 3–4
Elliptic filters, 18, 188–89
Elliptic lowpass filters, 101–12
  applications requirements, 106–10
  characteristics, 101–6
  general-purpose, 101–12
  pin descriptions for XR-1015/16, 110–12
Elliptical highpass filters, 35
EMI (electromagnetic interference), 175
Equal ripple response defined, 15
Error, sampling, 25
EXAR Corporation, 91, 101
Extended notch filters (LTC1062), 82
External
  clocks, 30, 71
  resistor values, 71–72

Fed-forward lowpass filters, 200
Feedback bandpass filter, multiple, 184
15-V supplies (LTC1062), IC filters operated from, 84
Filter octave tuning, lowpass, 88–90
Filter order, 3–4, 11
Filtering AC signals from high DC signals, 207
Filters
  active, 19–21
  allpass, 11
  approximations, 11–13
    attenuation rate near cutoff, 11
    filter order, 11
    monotonicity, 12
    passband ripple, 12
    rolloff rate, 11
    stopband ripple, 12
    transient response, 12
  attenuation of, 73
  band-reject, 7–8
  bandpass, 2, 5–7, 113–28
  cascading identical IC, 125–26
  cascading non-identical IC, 122–25
  choosing responses, 143–46
  clock-tunable notch, 84
  configurations, 31
  crystal, 191
  differences, 23–25
    accuracy, 23
    aliasing, 25
    circuit board area, 25
    component count, 25
    cost, 24
    design effort, 25
    frequency range, 24
    noise, 24
    offset voltage, 24
    tunability, 24–25
  electronic, 1–25
  elliptic, 188–89
  elliptical highpass, 35
  extended notch, 82

Filters (*Cont.*):
   first-order DC-accurate lowpass, 88
   full-duplex 300-baud modem, 44–45
   functions
      Bessel filters, 16–18
      Butterworth filters, 13
      Chebyshev filters, 13–16
      classic, 13–18
      elliptic filters, 18
      Thompson filters, 16–18
   high-cut, 9
   highpass, 9–11
   IC, 47
   infrasonic, 172–73
   input, 42–43
      impedances, 60
   low-cut, 10
   lowpass, 8–9, 65–100
   noise in, 146–48
   noises, 73
   notch, 7–8, 80–82
   order of, 3
   passive, 18–19
   performance of, 59–60
      filter input impedances, 60
      MAX270/271 noise, 60
   phase relationships of IC, 119–21
   phase shift in, 121
   phase-shift, 11
   Q, 4–5, 53
   simple 5-Hz, 82–83
   switched-capacitor, 19, 21–23
      IC, 27–45
   ultrasonic, 172–73
First trial values, 75
First-order DC-accurate lowpass filter (MAX280), 88
5-Hz, filters (LTC1062) simple, 82–83
5-pole lowpass filters, monolithic, 181–83
4.5-MHz notch filters, 201

Frequencies
   center, 5, 30–31
   corner, 41
   cutoff, 5, 41
   linear phase changes with, 39
   programming cutoff, 54–58
   ranges of, 24
   responses, 2
Frequency limits, passband, 5
Full-duplex 300-baud modem filter, 44–45
Functions
   network, 2
   transfer, 1–3, 2

Gain
   bandpass, 119–20
   and phase relationships of IC filters, 119–21
Generator, quadrature sine-wave, 43–44
Gyrator, notch filter using op amp as, 184

High-cut filters, 9
Highpass
   active RC filter, 157–59
   Butterworth filter, simple, 192–93
   filters, 9–11
      active RC, 195–96
      elliptical, 35
      with synthetic inductor, 174–76
      wideband, 199

IC (integrated circuit) filters, 47
   cascading identical, 125–26
   cascading non-identical, 122–25
   gain and phase relationships of, 119–21

with high input voltages
(LTC1062), 84
operated from 15-V supplies
(LTC1062), 84
with programmed cutoff
frequencies (LTC1062), 84–88
switched-capacitor, 27–45
Impedance
filter input, 60
scaling, 156–57
Inductor, highpass filter with
synthetic, 174–76
Infrasonic filters, 172–73
Input
filters, 42–43
offset problems, 135–38
Input impedance, filter, 60
Integrators, switched-capacitor, 27–29
Internal
buffers, 73
oscillators, 68–71

Lenk's Digital Handbook (Lenk),
57–58
Linear phase
change with frequency, 39
shift, 16
Linear Technology, 65, 129
tables, 113–19
Log scales, 4–5
Low-cut filters, 10
Low-offset, 12th-order Butterworth
lowpass filter (LTC1062), 80
Lowpass filters, 8–9, 65–100
DC-accurate Bessel, 205–6
active, 47–63
active RC, 159–61, 195–96
DC-coupled, 186–87
for anti-aliasing, 178–80
Bessel, 198
Butterworth active RC, 185–86
continuous, 47–63

dual-tracking 3-kHz, 204–5
elliptic, 101–12
applications requirements,
106–10
characteristics, 101–6
pin descriptions for XR-1015/16,
110–12
fed-forward, 200
general-purpose, 91–100
monolithic 5-pole, 181–83
octave tuning (MAX280), 88–90
pin descriptions for XR-1001/8,
99–100
response and applications, 91–99
simple, 198–99
using internal oscillators, 68–71
wideband, 199
without DC offset, 173–74
zero DC-error, 65–90
7th-order Sallen/Key (MAX280),
78
DC-accurate cascaded
(LTC1062), 78–79
anti-aliasing, 74
buffered cascade (LTC1062),
79–80
choosing capacitor values, 71–72
choosing external resistor values,
71–72
circuit functions for
MAX280/LTC1062, 65–68
filter attenuation, 73
filter noises, 73
first-order DC-accurate
(MAX280), 88
input voltage range for
MAX280, 72–73
internal buffers, 73
low-offset, 12th-order
Butterworth (LTC1062), 80
LTC1062 characteristics, 75
simplified design approaches
(MAX280/LTC1062), 75–90

Lowpass filters, zero DC-error (*Cont.*):
  single-supply 5th-order
    (MAX280), 77–78
  transient response for MAX280,
    73–74
  using clock divider ratios, 68
  using external clocks, 71
  using internal oscillators,
    68–71
Lowpass/bandpass filter, DC-accurate,
  201
LPO (lowpass output), 174
LTC1062
  characteristics, 75
  circuit functions for, 65–68
  manufactured by Linear
    Technology, 65
LTC1062 lowpass filters
  simplified design approaches,
    75–90
  DC-accurate cascaded lowpass
    filter, 78–79
  buffered cascade lowpass filter,
    79–80
  clock-sweepable
    pseudo-bandpass notch
    filter, 84
  clock-tunable notch filter, 84
  extended notch filter, 82
  IC filter with high input voltage,
    84
  IC filter operated from 15-V
    supplies, 84
  IC filter with programmed cutoff
    frequencies, 84–88
  low-offset, 12th-order
    Butterworth lowpass filter,
    80
  notch filter, 80
  selective clock-sweepable
    bandpass filter, 84
  simple 5-Hz filter, 82–83
LTC1063 output DC-offset, 183

MAX270 control interface, 58
MAX270/271
  basic circuit functions for, 47–54
  filter Q, 53
  op amp, 53
  passband, 53
  programmable capacitors, 51–53
  quiescent current and shutdown,
    53–54
  shutdown, 53–54
  T/H, 53
  manufactured by Maxim, 47
  noise, 60
MAX271
  control interface, 58–59
  multiplexed operation for, 60
MAX280
  circuit functions for, 65–68
  input voltage range for, 72–73
  lowpass filters
    7th-order Sallen/Key lowpass
      filters, 78
    first-order DC-accurate lowpass
      filter, 88
    lowpass filter octave tuning,
      88–90
    notch filters, 80–82
    simplified design approaches,
      75–90
    single-supply 5th order lowpass
      filters, 77–78
  manufactured by Maxim, 65
  notch filter, 80–82
  transient response for, 73–74
MAX291-97, switched-capacitor
  filters, 207–9
Maxim, 47, 65
Maximally-flat response defined, 13
Maximum passband ripple defined, 14
MF10
  circuit functions for, 27–31
    center frequencies, 30–31
    external clocks, 30

inputs/outputs, 29
switched-capacitor integrators,
  27–29
design examples with, 42–45
  full-duplex 300-baud modem
    filter, 44–45
  input filter, 42–43
  quadrature sine-wave generator,
    43–44
  sample/hold, 42–43
manufactured by National
  Semiconductor, 27
Microprocessors, programming
  without, 62
Minimum allowable attenuation
  defined, 14
Mode 1: notch, bandpass, and
  lowpass, 32–35
Mode 1, cascading bandpass filters in,
  126–27
Mode 1A: non-inverting bandpass,
  inverting bandpass, lowpass,
  32
Mode 2: notch, bandpass, and
  lowpass, 35–36
Mode 2, cascading bandpass filters in,
  127–28
Mode 3: highpass, bandpass, and
  lowpass, 36–37
Mode 3A: highpass, bandpass,
  lowpass, and notch, 37–39
Mode 4: allpass, bandpass, and
  lowpass, 39–40
Mode 5: complex zeros, bandpass,
  and lowpass, 40–41
Mode 6A: single pole, highpass, and
  lowpass, 41–42
Mode 6B: single pole lowpass
  (inverting and non-inverting), 42
Modem filters, full-duplex 300-baud,
  44–45
Modes of operation, design hints for
  all, 31–32

Monolithic 5-pole lowpass filter,
  181–83
Monotonicity, 12
Multiple-feedback
  active filters, 21
  bandpass filters, 184
Multiplexed operation for MAX271,
  60

National Semiconductor, 27
Networks
  crossover, 11
  functions, 2
  piezo-ceramic-based filter with
    differential, 190–91
Noises, 24
  filter, 73
  in filters, 146–48
  MAX270/271, 60
Norton amplifiers, 155–67
Notch filters, 7–8, 80–82
  4.5-MHz, 201
  DC-accurate, 176–78
  with adjustable Qs, 193
  bi-quad, 188
  clock-sweepable pseudo-bandpass,
    84
  clock-tunable, 84
  easily-tuned, 194
  extended, 82
  with high Qs, 196–97
  using op amps as gyrators, 184
Notch outputs, filters with, 184–85

Octave tuning (MAX280), lowpass
  filter, 88–90
Offset problems, input, 135–38
Offset voltage, 24
Op amp, 53
Op amp as gyrator, notch filter using,
  184

Oscillators, internal, 68–71
OTAs (operational transconductance amplifiers), 155

Passband, 53
  defined, 2
  frequency limits, 5
  ripple, 12, 14
Passive filters, 18–19
Phase change with frequency, linear, 39
Phase relationships of IC filters, 119–21
Phase response, 2
Phase shifts, 1–3, 2
  in filters, 121
  linear, 16
Phase-shifts, filters, 11
Piezo-ceramic-based filters
  basic, 190
  with differential networks, 190–91
Pin descriptions
  for XR-1001/8, 99–100
  for XR-1015/16, 110–12
PLLs (phase-locked loops),
    DC-accurate filters for, 169–71
Poles, 3–4
Power supplies
  configurations, 60–62
  spike suppressor for unregulated, 201
Programmable capacitors, 51–53
Programming
  cutoff frequency, 54–58
  without microprocessors, 62

Q
  bandpass filter with high, 203
  bandpass gain versus, 119–20
  filter, 4–5, 53
  notch filter with adjustable, 193
  notch filter with high, 196–97
  versus bandwidth, 121
Quadrature sine-wave generator, 43–44
Quiescent current and shutdown, 53–54

R5609 manufactured by Reticon, 101
Range, frequency, 24
RC filter circuits, biasing CFA active, 157
RC filters
  DC-coupled lowpass active, 186–87
  active, 155–67
    bandpass active RC filter, 161–63
    bandpass active RC filter with two CFAs, 163–65
    biasing CFA active RC filter circuits, 157
    capacitor characteristics, 156
    CFA operation, 155
    highpass active RC filter, 157–59
    impedance scaling, 156–57
    lowpass active RC filter, 159–61
    resistor characteristics, 156
    sensitivity functions, 157
    using CFAs (current-feedback amplifiers), 155–67
  bandpass active, 161–63
  basics, active, 155–57
  highpass active, 157–59
  lowpass active, 159–61
  lowpass Butterworth active, 185–86
  SCFs versus active, 129–30
  with three CFAs, 165–67
  with two CFAs, 163–65
RC highpass filters, active, 195–96
RC lowpass filters, active, 195–96
RC vs. SCF, active, 146–48
Resistors, external values, 71–72
Resistors characteristics, 156
Response, 1–3

Reticon, 101
Ringing, 12
Ripple, 6
  maximum passband, 14
  passband, 12
  response, 15
  stopband, 12
Rolloff rate, 11

Sallen-Key active filters, 20–21
Sampling error, 25
Scales, log, 4–5
Scaling, impedance, 156–57
SCFs (switched-capacitor filters), 129–53, 169
  active RC vs., 146–48
  aliasing in, 139–43
  bypass capacitors for, 150–53
  choosing filter response, 143–46
  circuit board layout problems, 130–32
  clock problems in, 148–50
    clock feedthrough, 150
    synchronized clocks, 150
  input offset problems, 135–38
  noise in filters, 146–48
  power supply problems, 132–35
  practical considerations for, 129–53
  slew limiting, 138–39
  versus active RC filters, 129–30
2nd-order filter section, single, 192
Sensitivity functions, 157
7th-order Sallen/Key lowpass filters (MAX280), 78
Shutdown, 53–54
Simplified Design of IC Amplifiers (Lenk), 2, 139, 155, 174–75
Simplified Design Microprocessor Supervisory Circuits (Lenk), 62
Simplified Design of Switching Power Supplies (Lenk), 133
SINAD (S/N plus distortion), 183

Sine-wave generator, quadrature, 43–44
Single-supply 5th order lowpass filter (MAX280), 77–78
Slew limiting, 138–39
Spike suppressor for unregulated power supplies, 201
Starting point, 19
State-space filter, universal, 187
Stopband ripple, 12
Stopbands defined, 6
Switched-capacitor filters, 19, 21–23, 207–9
Switched-capacitor IC filters, 27–45
  basic filter configurations, 31
  design examples with MF10, 42–45
  design hints for all modes of operation, 31–32
  mode 1: notch, bandpass, and lowpass, 32–35
  mode 1A: non-inverting bandpass, inverting bandpass, lowpass, 32
  mode 2: notch, bandpass, and lowpass, 35–36
  mode 3: highpass, bandpass, and lowpass, 36–37
  mode 3A: highpass, bandpass, lowpass, and notch, 37–39
  mode 4: allpass, bandpass, and lowpass, 39–40
  mode 5: complex zeros, bandpass, and lowpass, 40–41
  mode 6A: single pole, highpass, and lowpass, 41–42
  mode 6B: single pole lowpass (inverting and non-inverting), 42
  typical, 27–45
Switched-capacitor integrators, 27–29
Synthetic inductor, highpass filter with, 174–76

T/H (track-and-hold), 47, 53
Tables, using, 113–19
Thompson filters, 16–18
3 dB frequencies, 4–5
3-kHz lowpass filters, dual-tracking, 204–5
Threshold levels, digital, 59
Track-and-hold (T/H), MAX271, 60
Transfer functions, 1–3
   defined, 2
Transient response, 12
Trial values, first, 75
Tunability, 24–25
Tuned filter, two-stage, 194
Tuned notch filter, easily, 194
Tuned-filter circuit, basic, 194–95

Ultrasonic filters, 172–73
Universal state-space filter, 187
Universal state-variable active filters, 21
Unregulated power supplies, spike suppressor for, 201

Values, capacitor, 71–72
Voltages
   IC filters with high input, 84
   offset, 24

Wideband
   highpass filters, 199
   lowpass filters, 199
World, best filter in, 23–25

WTAs (wideband transconductance amplifiers), 174–75, 178

XR-1001/8
   manufactured by EXAR Corporation, 91
   pin descriptions for, 99–100
XR-1015/16
   manufactured by EXAR Corporation, 101
   pin descriptions for, 110–12

Zero DC-error lowpass filters, 65–90
   anti-aliasing, 74
   choosing
      capacitor values, 71–72
      external resistor values, 71–72
   circuit functions for MAX280/LTC 1062, 65–68
   filter
      attenuation, 73
      noises, 73
   input voltage range for MAX280, 72–73
   internal buffers, 73
   LTC1062 characteristics, 75
   simplified design approaches (MAX280/LTC1062), 75–90
   transient response for MAX280, 73–74
   using
      clock divider ratios, 68
      external clocks, 71
      internal oscillators, 68–71
Zeroes, 3–4